LO ON MATHE ATICAL SOCIETY ST

ng editor ssor W. Bruce, Department of Mathematics
ity of Li l, United Kingdom

3 *Local fields,* J. W. S. CASSELS
4 *An introduction to twister theory, second edition,* S. A. HUGGET & K. P. TOD
5 *Introduction to general relativity,* L. P. HUGHSTON & K. P. TOD
7 *The theory of evolution and dynamical systems,* J. HOFBAUER & K. SIGMUND
8 *Summing and nuclear norms in Banach space theory,* G. J. O. JAMESON
9 *Automorphisms of surfaces after Nielson and Thurston,* A. CASSON & S. BLEILER
11 *Spacetime and singularities,* G. NABER
12 *Undergraduate algebraic geometry,* M. REID
13 *An introduction to Hankel operators,* J. R. PARTINGTON
15 *Presentations of groups, second edition,* D. L. JOHNSON
17 *Aspects of quantum field theory in curved spacetime,* S. A. FULLING
18 *Braids and coverings: Selected topics,* V. LUNDSGAARD HANSEN
19 *Steps in commutative algebra,* R. Y. SHARP
20 *Communication theory,* C. M. GOLDIE & R. G. E. PINCH
21 *Representations of finite groups of Lie type,* F. DIGNE & J. MICHEL
22 *Designs, graphs, codes, and their links,* P. J. CAMERON & J. H. VAN LINT
23 *Complex algebraic curves,* F. KIRWAN
24 *Lectures on elliptic curves,* J. W. S. CASSELS
25 *Hyperbolic geometry,* B. IVERSEN
26 *An introduction to the theory of L-functions and Eisenstein series,* H. HIDA
27 *Hilbert space: Compact operators and the trace theorem,* J. R. RETHERFORD
28 *Potential theory in the complex lane,* T. RANSFORD
29 *Undergraduate cummutative algebra,* M. REID
31 *The Laplacian on a Riemannian manifold,* S. ROSENBERG
32 *Lectures on Lie groups and Lie algebras,* R. CARTER, G. SEGAL & I. MACDONALD
33 *A primer of algebraic* D*-modules,* S. C. COUTINHO
34 *Complex algebraic surfaces,* A. BEAUVILLE
35 *Young tableaux,* W. FULTON
37 *A mathematical introduction to wavelets,* P. WOJTASZCZYK
38 *Harmonic maps, loop groups and integrable systems,* M. GUEST
39 *Set theory for the working mathematician,* K. CIESIELSKI
40 *Ergodic theory and dynamical systems,* M. POLLICOTT & M. YURI
41 *The algorithmic resolution of diophantine equations,* N. P. SMART
42 *Equilibrium states in ergodic theory,* G. KELLER
43 *Fourier analysis on finite groups and applications,* A. TERRAS
44 *Classical invariant theory,* P. J. OLVER
45 *Permutation groups,* P. J. CAMERON
46 *Riemann surfaces: A primer,* A. BEARDON
47 *Introductory lectures on rings and modules,* J. BEACHY
48 *Set theory,* A. HAJNÁL & P. HAMBURGER
49 *An introduction to K-theory for* C**-algebras,* M. RØRDAM, F. LARSEN & N. LAUSTSEN
50 *A brief guide to algebraic number theory,* H. P. F. SWINNERTON-DYER
51 *Steps in commutative algebra,* R. Y. SHARP
52 *Finite Markov chains and algorithmic applications,* O. HÄGGSTRÖM

The Prime Number Theorem

G. J. O. JAMESON
Lancaster University

PUBLISHED BY THE PRESS SYNDICATE OF THE UNIVERSITY OF CAMBRIDGE
The Pitt Building, Trumpington Street, Cambridge, United Kingdom

CAMBRIDGE UNIVERSITY PRESS
The Edinburgh Building, Cambridge CB2 2RU, UK
40 West 20th Street, New York, NY 10011–4211, USA
477 Williamstown Road, Port Melbourne, VIC 3207, Australia
Ruiz de Alarcón 13, 28014 Madrid, Spain
Dock House, The Waterfront, Cape Town 8001, South Africa

http://www.cambridge.org

First published 2003

Printed in the United Kingdom at the University Press, Cambridge

A catalog record for this book is available from the British Library.

Library of Congress Cataloging in Publication data

Jameson, G. J. O. (Graham James Oscar)
The prime number theorem / G.J.O. Jameson.
p. cm. – (London Mathematical Society student texts ; 53)
Includes bibliographical references and index.
ISBN 0-521-81411-1 – ISBN 0-521-89110-8 (pb.)
1. Numbers, Prime. I. Title. II. Series

QA246 .J36 2003
512′.72–dc21 2002074199

ISBN 0 521 81411 1 hardback
ISBN 0 521 89110 8 paperback

Contents

Preface

How many prime numbers are there less than a given number n? Needless to say, an exact answer, in the form of an expression involving n, is not available. However, it is reasonable to ask whether there is a simple formula that *approximates* (in some sense) to the answer for all n. This is a very natural question to ask, but the prime numbers appear to be distributed in a very irregular way, and the reader will be in good company if he or she cannot imagine how to begin to answer it. However, it has been answered: the *prime number theorem* states that the required approximation is given by $n/(\log n)$ (and even better by other related formulae). This theorem is without question one of the really great theorems of mathematics. It is the central topic of this book.

Let us place the theorem in the mathematical landscape. It belongs, in principle, to analytic number theory, which is concerned with the estimation of number-theoretic quantities. While it is clearly one of the best theorems in this subject, it occupies a somewhat off-centre position there because of the fact that its most accessible proof requires hardly any number theory but quite a lot of analysis. For this reason, it does not fit very comfortably in books on analytic number theory, often appearing as an outlying topic. Meanwhile, if it is mentioned at all in a book on analysis, it will be in the role of a far-out application. Neither scenario does justice to such a great theorem! It is important and distinctive enough to be treated as a subject in its own right, rather than a fringe topic of either number theory or analysis. This is the rationale of the following pages. Our objective is to present an account in which the theorem, together with refinements, applications and results of a similar kind, is centre stage.

Mathematical problems are not usually solved by being considered in isolation. The reader may care to reflect on the combination of ideas that are used to evaluate the number π (in the sense of finding an infinite series converging to it). The proof of the prime number theorem is an unrivalled example of the harnessing of a number of seemingly unrelated ideas to achieve the desired result. As the machinery is assembled, it emerges that the original problem can be seen as one example of a more general problem. Strange as it may sound at this stage, our "fundamental theorem" will take the form of a statement about the coefficients of a certain kind of series called a "Dirichlet series", given facts about the function defined by the sum of the series. The prime number theorem will then appear as one case (admittedly the most prominent one) of the general theorem.

The topic is outstandingly suitable for the final year of an undergraduate programme in mathematics. The basic question is understandable from the outset and clearly worth asking. The solution takes students on a trip through some spectacular (but not excessively difficult) mathematical scenery. They will encounter, perhaps for the first time, such unexpected gems as the Euler product, the various Dirichlet series derived from it, the corresponding convolutions, Möbius inversion, the inversion of Dirichlet series by suitable integrals, and the extension of the zeta function by a clever rewriting of the defining formula. They will also meet the Riemann hypothesis, commonly regarded as the most important unsolved problem in mathematics.

As already indicated, the main prerequisites lie in analysis rather than number theory. More exactly, we presuppose standard undergraduate courses in both real and complex analysis, including, for example, Riemann integration, uniform convergence and Cauchy's integral theorem. It is hoped that readers will appreciate seeing how the theorems of analysis, which they may have regarded as being very theoretical, can be applied to a "real" problem – furthermore, to one for which their relevance would hardly have been suspected. Some care is taken to avoid dependence on concepts that are not really needed: in particular, no knowledge of Lebesgue integration is required. A number of topics on the borderline of what might be included in a first course on analysis, such as infinite products and multiplication of series, are treated in appendices.

Certain sections presuppose a very basic knowledge of number theory, but for the core proof of the prime number theorem even this is not really necessary.

We make no attempt to contrast analysis and number theory, or to keep them separate. Indeed, part of the beauty of this subject is the way in which ideas from the two strands weave together and operate in partnership. A typical benefit is the avoidance of unmotivated definitions: for example, we can allow the Euler product to "discover" the Möbius and von Mangoldt functions for us, rather than asking the reader to accept strange-seeming definitions out of the blue.

Our basic account of the prime number theorem is contained in the first three chapters. As already remarked, a wide assortment of ideas is encountered on the way. The aim is to develop each topic far enough to give the reader an adequate feeling for them; sometimes this means going a little beyond what is strictly needed for the ultimate purpose. Also, as stated, our agenda stretches to include "results of a similar kind". As an example, consider the sum of $1/p$ for all primes $p \leq n$; the original problem is similar, but with 1 instead of $1/p$. It happens that the quantity just stated can be estimated much more easily, so our account includes this estimation even though it is not actually needed to prove the prime number theorem. Several other estimations of number-theoretic functions (more exactly, their partial sums) are included in the same spirit. In this way, the book does in fact cover a certain amount of the core material of analytic number theory, without in any way aiming to give a comprehensive treatment of the subject. For the reader wanting a minimal path to the main theorem, all such deviations are identified.

In chapter 3, two alternative methods are given for the final derivation of the fundamental theorem. One is a new variant of the traditional method based on Mellin (or Perron) inversion of Dirichlet series. The other is Newman's method, which has attracted considerable attention since its appearance in 1980. Readers can judge for themselves the relative merits of the two methods. Either method leads to a prior version of the theorem taking the form of an integral identity. From it one can simultaneously deduce *limit* results (such as the prime number theorem itself) and the corresponding *series* results that are also of considerable interest. This differs from the accounts found in most, if not all, existing books, which require a further stage of work to deduce the series versions from the limit ones (or conversely).

Having at length established the prime number theorem, we describe some applications and extensions, such as the estimation of the number of integers less than n having k prime factors.

The remaining three chapters continue the study further in three mutually independent directions. Chapter 4 is concerned with Dirichlet's theorem on the equal distribution of prime numbers among residue classes. It is beautiful to see how the notion of *Dirichlet characters* enables us to adapt the original method for this purpose, with L-functions taking the place of the zeta function. This chapter requires a minimal knowledge of group theory.

No statement about approximation is complete without a discussion of accuracy. Chapter 5 addresses the problem of error estimates in the prime number theorem and similar results. The Mellin inversion method adapts in a straightforward way to provide the classical error estimate (Newman's proof is less suitable for this purpose). Readers need to get this far to appreciate the significance of the Riemann hypothesis, which can now be presented as being equivalent to the much stronger error estimates that seem to be supported by numerical evidence.

Finally, chapter 6 contains an account of one version of the "elementary" proof of the prime number theorem by essentially number-theoretic methods, avoiding complex analysis.

Proofs and explanations are given at a level of detail suitable for final-year undergraduate students. Results (and only results) have a three-point numbering: $m.n.p$ is result p of section $m.n$. The most important results are labelled "theorems", and all others "propositions", unless they are lemmas or corollaries, but they are usually referred to simply by the number. Exercises serve the usual dual purpose of giving the reader essential practice and of supplementing the information included in the text.

Most of the number-theoretic estimations are accompanied by numerical tables, which some readers may wish to check or supplement by their own computer calculations. These tables serve to put some "flesh" on the results presented and to illustrate the enormous variation between the rates of convergence in different cases.

To avert possible disappointment, it would be fair to mention one topic that is *not* included, the computation of very large prime numbers. This is currently a popular occupation, and it certainly utilises some interesting results of number theory. However, it has little relevance to the prime number theorem, so the reader who wants it must be directed elsewhere.

Acknowledgements. Special acknowledgement is due to my son, Timothy Jameson, for awakening my interest in this subject in the first place, for numerous comments and suggestions and for contributing the computational elements, including most of the numerical tables and the sketch of the zeta function in section 3.1. I would also like to thank Gordon Blower, Robin Chapman and Ross Lawther for comments, corrections and suggestions, and Alison Woollatt and Elise Oranges of Cambridge University Press for their patient help and advice in achieving a very pleasing layout.

1
Foundations

In this chapter, we assemble a number of ideas and techniques that will eventually be fitted together to achieve our aim. Their only common feature is that they are needed to prove the prime number theorem, so the chapter has no single unifying theme. However, each *section* of the chapter is devoted to a very clearly defined topic. Some of these ideas are analytic, others number-theoretic, but there would be no advantage in trying to keep the two strands apart: they reinforce each other in a fruitful partnership.

Our objective is to find a formula that approximates $\pi(x)$, the number of primes not greater than x. We start, in section 1.1, by identifying some candidates as suggested by numerical evidence. We also give a brief account of the long history leading to the successful proof of the prime number theorem.

The term "arithmetic function" is used for a sequence defined using number-theoretic properties in some way. A great deal of number theory consists of the study of such functions. Now we can express $\pi(x)$ as the partial sum $\sum_{n\leq x} u_P(n)$, where u_P is the arithmetic function defined as follows:

$$u_P(n) = \begin{cases} 1 & \text{if } n \text{ is prime,} \\ 0 & \text{otherwise.} \end{cases}$$

Typically, arithmetic functions appear to be very irregular, but this is smoothed out by addition, and one can hope to find an estimate for their partial sums. This identifies our problem as one of a certain type.

We go on to describe two essential techniques for rewriting and estimating discrete sums, Abel summation and integral estimation. Both are used constantly in all that follows.

After this, we are in a position to describe the first real progress towards the prime number theorem, achieved by Chebyshev in 1850. Chebyshev recognised that an estimation of $\pi(x)$ can be deduced from an estimation of $\theta(x) = \sum_{p\in P[x]} \log p$ (where $P[x]$ denotes the set of primes not greater than

x), and showed that the latter sum can be estimated by a comparatively short (but ingenious) argument. In this way, he demonstrated that $\pi(x)$ lies between $cx/\log x$ and $Cx/\log x$ for two constants c, C.

Finally, we introduce a concept that will permeate the rest of our study to the extent that it could serve as a subtitle for this book. A *Dirichlet series* is a series of the form $\sum_{n=1}^{\infty} a(n)/n^s$, in which s is a complex variable. The case $a(n) = 1$ defines the *Riemann zeta function.* Every arithmetic function has a corresponding Dirichlet series; multiplication of the series corresponds to "convolution" of the arithmetic functions. Our "fundamental theorems" in chapter 3 will derive information about the partial sums of $a(n)$ from the nature of the function defined by the series, with the prime number theorem appearing as a special case.

1.1 Counting prime numbers

As the reader surely knows, prime numbers are those that have no positive divisors except 1 and the number itself. The special significance of prime numbers is due to the following fact, which we will assume known:

Every positive integer is expressible as a product of primes. The expression is unique if the primes are listed in increasing order.

Effectively, this means that the primes are the basic "atoms" in the multipicative system of integers. (If n is itself prime, we are regarding it as a "product" of one prime, itself.)

The first result on the number of primes was already known to Euclid. Here it is, with Euclid's beautiful proof.

Proposition 1.1.1 *There are infinitely many prime numbers.*

Proof Choose finitely many primes $p_1, p_2, \ldots, p_n$. We will show that they cannot constitute the total set of primes. Consider the number

$$N = p_1 p_2 \ldots p_n + 1.$$

Then N is not a multiple of any p_j, because it clearly leaves remainder 1 when divided by p_j. However, by the above statement, N is expressible as a product of primes. Let q be any one of these. Then q is a further prime, different from all the p_j, which therefore indeed fail to constitute the total set of primes. □

Note This reasoning actually shows a bit more: if the primes are listed as $p_1, p_2, \ldots$ in increasing order, then $p_{n+1} \leq p_1 p_2 \ldots p_n + 1$.

With this settled, it is natural to ask how many prime numbers there are up to any given number. This is the topic of our study. Let us give it some notation:

$$\pi(x) = \text{the number of primes not greater than } x.$$

This notation is standard in number theory; there is no real danger of confusion with the number π. It will suit our purposes to regard $\pi(x)$ as a function of a real variable x. As such a function, it is, of course, constant between primes and jumps by 1 at each prime.

The first impression given by the sequence of primes

$$2, 3, 5, 7, 11, \ldots, 101, 103, 107, 109, 113, 127, \ldots, 163, 167, 173, 179, \ldots$$

is one of extreme irregularity. There are bunches, gaps and relatively uniform stretches. It would appear to be a daunting task to find a simple expression that approximates to $\pi(x)$ for all large enough x. The only simple observation is that the primes tend to become more sparse as one goes on. However, an examination of the numerical values of $\pi(x)$ suggests that a reasonable approximation is given by $x/(\log x)$, a considerably better one by

$$\frac{x}{\log x - 1},$$

and a still better one by the "logarithmic integral", defined as follows:

$$\mathrm{li}(x) = \int_2^x \frac{1}{\log t}\, dt.$$

Some of these numerical values are as follows (given to the nearest integer):

n	$\pi(n)$	$\frac{n}{\log n}$	$\frac{n}{\log n - 1}$	$\mathrm{li}(n)$
1,000	168	145	169	177
10,000	1,229	1,086	1,218	1,246
50,000	5,133	4,621	5,092	5,166
100,000	9,592	8,686	9,512	9,630
500,000	41,538	38,103	41,246	41,607
1,000,000	78,498	72,382	78,031	78,628
10,000,000	664,579	620,421	661,459	664,918

By the year 1800, long before the age of computers, mathematicians had performed the remarkable feat of calculating these figures by hand up to

$n = 400,000$. In the age of computers, it has of course become much easier to calculate values of $\pi(x)$. Some readers will be interested in doing so on their own computer: various methods for this are discussed in appendix F.

Let us formulate precisely the conjecture suggested by these figures. Given two functions $f(x)$, $g(x)$, both tending to infinity as $x \to \infty$, we write

$$f(x) \sim g(x) \qquad \text{as } x \to \infty$$

to mean that

$$\frac{f(x)}{g(x)} \to 1 \qquad \text{as } x \to \infty.$$

Our conjecture is the statement

$$\pi(x) \sim \text{li}(x) \qquad \text{as } x \to \infty.$$

In fact, as we will show in section 1.5,

$$\frac{x}{\log x} \sim \frac{x}{\log x - 1} \sim \text{li}(x) \qquad \text{as } x \to \infty,$$

so at this level it is equivalent to state the conjecture using any of the three functions.

The conjecture is in fact true. The statement $\pi(x) \sim \text{li}(x)$ is called the *prime number theorem*. It is indisputably one of the most celebrated theorems in mathematics. Ways of proving it, together with related results, more precise versions and generalizations, form the subject of this book. Of course, numerical evidence of the above type can never constitute a proof of the general statement.

An informal interpretation of the theorem is that the "average density" of primes around a large number x approximates to $1/(\log x)$, or that the "probability" of n being prime is (in some sense) $1/(\log n)$.

Let us return to the historical trail (for a much more detailed historical account, see [Nar]). Legendre, in 1798, postulated the approximations $x/(\log x)$ and $x/(\log x - 1)$. He suggested (wrongly) that an even better approximation would be given by $x/(\log x - A)$, with $A = 1.0836$. Meanwhile, Gauss proposed $\text{li}(x)$. It seems that Gauss recorded his conjecture around 1793 (at the age of 14!) but did not communicate it to anyone until 1849.

The search for a proof remained one of the main areas of mathematical endeavour during the rest of the nineteenth century. In 1850, a giant stride was made by Chebyshev, who showed, by essentially number-theoretic methods, that there are constants c and C (not very far from 1) such that

$$c\,\text{li}(x) \le \pi(x) \le C\,\text{li}(x)$$

for all large enough x. However, no refinement of his methods seemed to offer any hope of proving the desired limit.

A completely different approach was proposed by Riemann in 1859. His starting point was a remarkable identity already discovered by Euler in 1737, expressing the "zeta function"

$$\zeta(s) = \sum_{n=1}^{\infty} \frac{1}{n^s}$$

as an infinite *product* involving the primes. Riemann considered this as a function of a *complex* variable s, defined by the above formula for $\operatorname{Re} s > 1$. He showed how to extend the definition of the function to the rest of the complex plane and outlined a programme showing how, if certain properties of the extended zeta function could be established, the prime number theorem would follow. His paper was a bold imaginitive leap; it was hardly an obvious idea to use the theorems of complex analysis to count prime numbers! However, Riemann was not able to justify all his steps, and one of them, the "Riemann hypothesis", has remained unsolved to this day, regarded by many as the most important unsolved problem in mathematics.

It was not until 1896 that Riemann's programme was successfully completed. It was then done so independently by the French mathematician Jacques Hadamard and the Belgian Charles-Jean de la Vallée Poussin. They were able to bypass the Riemann hypothesis and establish other properties of the zeta function that were sufficient for the purpose. Hadamard lived until 1963 (aged 97) and de la Vallée Poussin until 1962 (aged 95): their mathematical labours cannot have done any harm to their health!

Further variations and modifications of their methods were developed by Mertens, Landau and others, but to this day the simplest, and most powerful, proofs of the prime number theorem rely on the zeta function and complex analysis, as suggested by Riemann. In chapters 1–3, we present a version (and a variant) that benefits from a century of "tidying up", but which still recognisably owes its existence to Riemann.

After the successful outcome of Riemann's programme, it remained a matter of great interest to ask whether the theorem could after all be proved by number-theoretic methods, without complex analysis. This was eventually achieved in 1949, again independently by two people, A. Selberg and P. Erdös. Proofs of this sort are called "elementary" as opposed to "analytic". However, "elementary" does not mean "simple"! Half a century later, known proofs of this sort are still more complicated than analytic ones and are less successful in providing error estimates or in delivering other theorems of the same sort. A version is presented in chapter 6.

Exercises

1 Let $n \geq 1$ and let $E = \{30n + r : 0 \leq r \leq 29\}$. For which values of r is $30n + r$ not a multiple of 2, 3 or 5 ? By considering the possible positions of multiples of 7, show that E contains at most seven primes (seven cases, no short cuts!).

2 Show that, for any $n \geq 2$, there is eventually a gap of length at least n between successive primes. [Hint: Consider $n!+2$ or $p_1 p_2 \dots p_n + 2$.]

3 Let the primes be listed, in order, as $p_1, p_2, \dots$. Use Euclid's proof to show by induction that $p_n \leq 2^{2^{n-1}}$ for each n. Deduce that

$$\pi(x) \geq \frac{\log \log x}{\log 2}.$$

1.2 Arithmetic functions

Formally, an arithmetic function is simply a sequence, with real or complex values. A sequence is, of course, a *function* on the set $\mathbb{N}$ of positive integers. To emphasize that we are thinking of them as functions, we shall usually use notation like $a(n)$, rather than a_n, for the value corresponding to the integer n. The term "arithmetic function" is used especially when $a(n)$ is defined using number-theoretic properties in some way. A large part of number theory consists, in one way or another, of the study of these functions.

We list some examples. First, two very simple ones, mainly to establish the notation we will use:

$$u(n) = 1 \quad \text{for all } n \quad (\text{the "unit function"});$$

$$e_j(n) = \begin{cases} 1 & \text{if } n = j, \\ 0 & \text{if } n \neq j. \end{cases}$$

Next, given any subset E of $\mathbb{N}$ (for example, the set P of primes), define

$$u_E(n) = \begin{cases} 1 & \text{if } n \in E, \\ 0 & \text{if } n \notin E. \end{cases}$$

Clearly, u itself is the case $E = \mathbb{N}$. Third, three more obviously number-theoretic examples:

$\tau(n) =$ the number of (positive) divisors of n, including 1 and n;

$\omega(n) =$ the number of prime divisors of n;

$\Omega(n) =$ the number of prime factors of n, counted with repetitions.

(This notation is more or less standard, though some writers use d instead of τ.) Note that $\tau(1) = 1$ and $\omega(1) = \Omega(1) = 0$. For $n > 1$, these functions can easily be described in terms of the prime factorization of n, as follows.

Proposition 1.2.1 *Suppose that $n > 1$, with prime factorization*

$$n = \prod_{j=1}^{m} p_j^{k_j}.$$

Then

$$\tau(n) = \prod_{j=1}^{m} (k_j + 1), \qquad \omega(n) = m, \qquad \Omega(n) = \sum_{j=1}^{m} k_j.$$

Proof The expressions for $\omega(n)$ and $\Omega(n)$ are just the definition. Divisors of n are of the form $\prod_{j=1}^{m} p_j^{r_j}$, where, for each j, the possible values of r_j are $0, 1, \ldots, k_j$. This gives the expression for $\tau(n)$. □

In particular, if p is prime, then $\tau(p^k) = k + 1$, $\omega(p^k) = 1$ and $\Omega(p^k) = k$. To give another example, since $72 = 2^3.3^2$, we have $\tau(72) = 12$, $\omega(72) = 2$ and $\Omega(72) = 5$.

Given arithmetic functions a,b, we denote the pointwise product by ab, so that $(ab)(n) = a(n)b(n)$. Obviously, $au = a$ for any a.

Summation functions. Given an arithmetic function $a(n)$, its *summation function* $A(x)$ is defined by

$$A(x) = \sum_{n \le x} a(n).$$

It is useful to regard $A(x)$ as a function of a *real* variable x. As such a function, it is, of course, constant between integers and has a jump discontinuity at each integer where $a(n) \neq 0$. Clearly, $\pi(x)$ is the summation function of $u_P(n)$ (note: in particular cases, the established notation will not usually allow a notational device like the substitution of A for a).

Individual values of arithmetic functions may fluctuate wildly – as in most of the examples just given. However, in many cases summation smooths out the fluctuation, and it may be possible to find an asymptotic expression for the summation function for large x. In the case of $\tau(n)$ and $\omega(n)$, the first step is to apply a bit of "lateral thinking" to obtain alternative expressions for the summation functions, as in the next result. The notation $[x]$ means the largest integer not greater than x. We shall use the notation $P[x]$ to mean the set of primes not greater than x (but there is no generally agreed notation for this). Also, $j|n$ means that j divides into n.

Proposition 1.2.2 *Write* $S_\tau(x) = \sum_{n \le x} \tau(n)$ *and* $S_\omega(x) = \sum_{n \le x} \omega(n)$. *Then*

$$S_\tau(x) = \sum_{j \le x} \left[\frac{x}{j}\right], \qquad S_\omega(x) = \sum_{p \in P[x]} \left[\frac{x}{p}\right].$$

Proof Clearly, $S_\tau(x)$ is the number of ordered pairs (j, n) such that $j|n$ and $n \le x$. For fixed j (instead of fixed n), the number of such pairs is the number of multiples rj not greater than x. This number is obviously $[x/j]$. The stated expression follows.

The argument for $S_\omega(x)$ is similar, counting the number of pairs (p, n) as above, but with p prime. □

The "double counting" principle seen in this proof is often useful. We shall see later how the identities in 1.2.2 can be used to derive asymptotic expressions for $S_\tau(x)$ and $S_\omega(x)$, and similar results will be obtained for some other arithmetic functions. Since $\pi(x)$ is itself a summation function, our main objective, the prime number theorem, is a result of exactly this type. But, as the reader may suspect, this case will cost us a lot more effort than most of the others.

Multiplicative functions. We denote by (m, n) the greatest common divisor of m and n. An arithmetic function a is said to be

completely multiplicative if $a(mn) = a(m)a(n)$ for all m, n;

multiplicative if $a(mn) = a(m)a(n)$ whenever $(m, n) = 1$.

Clearly, if a is multiplicative and not identically zero, then $a(1) = 1$. Also, a is fully determined by the values $a(p^k)$ for prime p, since if the prime factorization of n is $\prod_{j=1}^r p_j^{k_j}$, then $a(n) = \prod_{j=1}^r a(p_j^{k_j})$. Of course, if a is completely multiplicative, then $a(p^k) = a(p)^k$, and the function is already fully determined by the values $a(p)$.

We list some examples.

(i) For any s, let $a(n) = n^s$. Then a is completely multiplicative.

(ii) e_1 is completely multiplicative, but e_j is not multiplicative for $j \ge 2$, since $e_j(1) = 0$.

(iii) τ is multiplicative. This follows from 1.2.1, since if $(m, n) = 1$, then m and n have different prime divisors. It is not completely multiplicative, since $\tau(2) = 2$, $\tau(4) = 3$.

(iv) u_P is not multiplicative, since $u_P(1) = 0$.

(v) Neither ω nor Ω is multiplicative; however, we have $\Omega(mn) = \Omega(m) + \Omega(n)$, and similarly for ω when $(m, n) = 1$.

(vi) *Liouville's function* is defined by $\lambda(n) = (-1)^{\Omega(n)}$. It is completely multiplicative, by the statement in (v).

(vii) Let

$$\chi(n) = \begin{cases} 0 & \text{if } n \text{ is even} \\ 1 & \text{if } n \equiv 1 \pmod 4 \\ -1 & \text{if } n \equiv -1 \pmod 4). \end{cases}$$

By considering the different cases, one checks easily that χ is completely multiplicative.

As the reader may already know, there are many further interesting arithmetic functions. Some will make their appearance in later sections.

Exercises

1. Find the smallest n such that: (i) $\Omega(n) = 4$, (ii) $\omega(n) = 4$, (iii) $\tau(n) = 4$.
2. Show that $\sum_{n=1}^{30} \tau(n) = 111$ without calculating individual values of $\tau(n)$.
3. Calculate $\sum_{n=1}^{100} \omega(n)$ without calculating individual values of $\omega(n)$.
4. Show that $\tau(n)$ is odd if and only if n is a square.
5. Show that, for any $n \geq 2$,

$$2^{\omega(n)} \leq \tau(n) \leq 2^{\Omega(n)} \leq n.$$

   [You may assume that $k + 1 \leq 2^k$ for all $k \geq 0$.]

6. Let S be the set of squares. Show that u_S is multiplicative.
7. Let $a(n) = (-1)^{n-1}$ for $n \geq 1$. Show that a is multiplicative.
8. Prove that, for any $\varepsilon > 0$, we have $\tau(n)/n^{\varepsilon} \to 0$ as $n \to \infty$. [Again use $k+1 \leq 2^k$. For each prime $p < 2^{1/\varepsilon}$, show that there is a constant C_p such that $k + 1 \leq C_p\, p^{\varepsilon k}$ for all k.]

1.3 Abel summation

Discrete version

Abel summation, in its various forms, will be a very basic tool in all that follows, so we will describe it rather thoroughly. It is the process of expressing a sum of products $\sum a(r)f(r)$ in terms of *partial sums* of the $a(r)$'s and *differences* of the $f(r)$'s. Our choice of notation reflects the different roles played by $a(r)$ and $f(r)$. The process is exactly analogous to integration by parts for functions, and indeed it is sometimes called "summation by parts" or "partial summation". As already mentioned, a central theme in analytic number theory is the estimation of partial sums (rather than individual values) of arithmetic functions. This explains why Abel summation is so often appropriate. Throughout the following, we assume that $a(r)$, $f(r)$ are given numbers (real or complex) for $r \geq 1$, and write $A(n) = \sum_{r=1}^{n} a(r)$ for $n \geq 1$ (also $A(0) = 0$). If we have another sequence $b(r)$, then $B(n)$ is defined similarly. The basic result is very simple, as follows.

Proposition 1.3.1 *For integers $n > m \geq 0$,*

$$\sum_{r=m+1}^{n} a(r)f(r) = \sum_{r=m}^{n-1} A(r)[f(r) - f(r+1)] + A(n)f(n) - A(m)f(m).$$

In particular,

$$\sum_{r=1}^{n} a(r)f(r) = \sum_{r=1}^{n-1} A(r)[f(r) - f(r+1)] + A(n)f(n).$$

Proof The proof looks nicer if we write A_r and f_r instead of $A(r)$ and $f(r)$! For all $r \geq 1$, we have $a_r = A_r - A_{r-1}$ (recall $A_0 = 0$). Hence

$$\begin{aligned}\sum_{r=m+1}^{n} a_r f_r &= (A_{m+1} - A_m)f_{m+1} + (A_{m+2} - A_{m+1})f_{m+2} + \cdots \\ &\quad \cdots + (A_n - A_{n-1})f_n \\ &= -A_m f_{m+1} + \sum_{r=m+1}^{n-1} A_r(f_r - f_{r+1}) + A_n f_n \qquad (1.1) \\ &= -A_m f_m + \sum_{r=m}^{n-1} A_r(f_r - f_{r+1}) + A_n f_n. \qquad (1.2)\end{aligned}$$

The second statement is the case $m = 0$. □

This simple identity has numerous corollaries and applications.

Corollary 1.3.2 *Suppose that $f(r)$ is real and non-negative, and decreases with r. Suppose that $a(r)$, $b(r)$ are such that $A(r) \leq CB(r)$ for all r. Then*

$$\sum_{r=1}^{n} a(r)f(r) \leq C \sum_{r=1}^{n} b(r)f(r).$$

Proof This follows at once from 1.3.1, because $f(r) - f(r+1)$ and $f(n)$ are non-negative. □

Note Taking $a(r) = 1$ in 1.3.1, we obtain an expression for $\sum_{r=1}^{n} f(r)$ itself:

$$\sum_{r=1}^{n} f(r) = \sum_{r=1}^{n-1} r[f(r) - f(r+1)] + nf(n).$$

Next, we describe some applications to infinite series.

Proposition 1.3.3 *Suppose that $A(n)f(n) \to 0$ as $n \to \infty$. Then if one of the series*

$$\sum_{r=1}^{\infty} a(r)f(r) \qquad \textit{and} \qquad \sum_{r=1}^{\infty} A(r)[f(r) - f(r+1)]$$

converges, then so does the other, to the same sum.

Proof Recall that the sum of a series means the limit as $n \to \infty$ of the sum of n terms. Given this, the statement follows at once from 1.3.1. □

Note that by taking limits in the first statement in 1.3.1, we have also

$$\sum_{r=m+1}^{\infty} a(r)f(r) = \sum_{r=m}^{\infty} A(r)[f(r) - f(r+1)] - A(m)f(m).$$

Proposition 1.3.4 (Dirichlet's test for convergence) *Suppose that:*

(i) $|A(r)| \leq C$ *for all* r,

(ii) $f(r) \to 0$ *as* $r \to \infty$,

(iii) $\sum_{r=1}^{\infty} |f(r) - f(r+1)|$ *is convergent.*

Then $\sum_{r=1}^{\infty} a(r)f(r)$ *converges, say to* S, *where*

$$|S| \leq C \sum_{r=1}^{\infty} |f(r) - f(r+1)|.$$

Condition (iii) can be replaced by: (iiia) $f(r)$ is non-negative and decreasing. We then have $|S| \leq Cf(1)$.

Proof By the comparison test for series, $\sum_{r=1}^{\infty} A(r)[f(r) - f(r+1)]$ is convergent, with the sum S satisfying the stated inequality. Also, clearly, $A(r)f(r) \to 0$ as $r \to \infty$. The statement follows, by 1.3.3. Under condition (iiia), we have

$$\sum_{r=1}^{n} |f(r) - f(r+1)| = \sum_{r=1}^{n} [f(r) - f(r+1] = f(1) - f(n+1),$$

so that $\sum_{r=1}^{\infty} |f(r) - f(r+1)| = f(1)$. □

Note that, under the conditions of this result, $\sum_{r=1}^{\infty} A(r)\,[f(r)-f(r+1)]$ is *absolutely* convergent, while $\sum_{r=1}^{\infty} a(r)f(r)$ may well not be.

In particular, condition (i) implies that, for all real $s > 0$, the series $\sum a(r)/r^s$ converges, say to S, where $|S| \leq C$. Later, we shall be considering series of this type for *complex* s; these are more easily handled by the continuous version of Abel summation, discussed below.

Abel summation is often used to give estimations of partial sums, given information about other partial sums. The next example illustrates this.

Example 1 Given that $|\sum_{r=1}^{n} a(r)/r| \leq C$ for all n, to show that $|A(n)| \leq (2n-1)C$ for all n.

Write $a(r)/r = b(r)$. Then we have $|B(n)| \leq C$ for all n and, by 1.3.1,

$$A(n) = \sum_{r=1}^{n} rb(r) = -B(1) - B(2) - \cdots - B(n-1) + nB(n).$$

The stated inequality is now clear.

Continuous version

Given an arithmetic function $a(r)$, its summation function $A(x)$ is defined for all positive, real x, not just integers. Meanwhile, in products of the type we are considering, the numbers $f(r)$ (as our notation already suggests) are often the values at integers of a function $f(x)$ of a real variable. When this happens, we can replace the discrete sums in the above expressions by integrals, using the elementary fact that $f(r+1) - f(r) = \int_r^{r+1} f'(t)\,dt$ (this remains valid when f is complex-valued: for clarification, see appendix A). The resulting identity, in its various forms, is sometimes called *Abel's summation formula.* It is very useful, because integrals are often easier to evaluate than discrete sums.

Throughout the following, the numbers $a(r)$ (for integers $r \geq 1$) are assumed given, and $A(x) = \sum_{r \leq x} a(r)$. We start by giving the continuous equivalent of the first expression in 1.3.1. The similarity with integration by parts is even more apparent.

Theorem 1.3.5 *Let $y < x$, and let f be a function (with real or complex values) having a continuous derivative on $[y, x]$. Then*

$$\sum_{y<r\leq x} a(r)f(r) = A(x)f(x) - A(y)f(y) - \int_y^x A(t)f'(t)\,dt.$$

Proof Let m, n be integers such that $n \le x < n+1$ and $m \le y < m+1$. Then (treating the symbols $<$ and $\le$ with due respect) we have

$$\sum_{y<r\le x} a(r)f(r) = \sum_{r=m+1}^{n} a(r)f(r).$$

We use identity (1.1) from the proof of 1.3.1. Since $A(t) = A(r)$ for $r \le t < r+1$, we have

$$\begin{aligned}\sum_{r=m+1}^{n-1} A(r)[f(r) - f(r+1)] &= -\sum_{r=m+1}^{n-1} A(r)\int_r^{r+1} f'(t)\,dt \\ &= -\sum_{r=m+1}^{n-1} \int_r^{r+1} A(t)f'(t)\,dt \\ &= -\int_{m+1}^{n} A(t)f'(t)\,dt.\end{aligned}$$

(Note that the integral of $A(t)f'(t)$ on $[r, r+1]$ is not affected by the possibly new value of $A(t)$ at the single point $r+1$.) Also, $A(t) = A(n)$ for $n \le t \le x$, so

$$A(x)f(x) - A(n)f(n) = A(n)[f(x) - f(n)] = \int_n^x A(t)f'(t)\,dt,$$

and similarly

$$A(m)f(m+1) - A(y)f(y) = A(m)[f(m+1) - f(y)] = \int_y^{m+1} A(t)f'(t)\,dt.$$

Using these identities to substitute for $A(m)f(m+1)$ and $A(n)f(n)$ in (1.1), we obtain the stated identity. □

Most of the time we shall apply the formula to sums starting at $r = 1$. Two variant forms for this case are given in the next result.

Proposition 1.3.6 *Let f have a continuous derivative on $[1, x]$. Then:*

(i) $$\sum_{r\le x} a(r)f(r) = A(x)f(x) - \int_1^x A(t)f'(t)\,dt,$$

(ii) $$\sum_{r\le x} a(r)[f(x) - f(r)] = \int_1^x A(t)f'(t)\,dt.$$

Proof (i) Take $y = 1$ in 1.3.5. Then $A(y)f(y) = a(1)f(1)$, while the sum on the left is $\sum_{2\le r\le x} a(r)f(r)$. When combined, this gives the required sum.

(ii) This follows, because $\sum_{r\le x} a(r)f(x) = A(x)f(x)$. □

Sometimes we will be interested in a function f that is undefined at 1 (typically $f(t) = 1/\log t$). The following variant of the formula deals with this situation.

Proposition 1.3.7 *If f has a continuous derivative on $[2, x]$ and $a(1) = 0$, then*

$$\sum_{2 \le r \le x} a(r)f(r) = A(x)f(x) - \int_2^x A(t)f'(t)\,dt.$$

Proof Taking $y = 2$ in 1.3.5, we have

$$\begin{aligned}\sum_{2\le r\le x} a(r)f(r) &= a(2)f(2) + \sum_{2<r\le x} a(r)f(r)\\ &= a(2)f(2) + A(x)f(x) - A(2)f(2) - \int_2^x A(t)f'(t)\,dt.\end{aligned}$$

Since $a(1) = 0$, we have $A(2) = a(2)$, so the statement follows. □

To assist in the process of getting used to these formulae, we list a number of particular cases (with general $a(r)$ but specific $f(t)$). We remark that when similar expressions appear in later sections, the notation will often be n instead of r. The reader may like to write out and compare some of the corresponding discrete expressions.

(i) $\displaystyle\sum_{r\le x} \frac{a(r)}{r} = \frac{A(x)}{x} + \int_1^x \frac{A(t)}{t^2}\,dt;$

(ii) $\displaystyle\sum_{y<r\le x} a(r)\log r = A(x)\log x - A(y)\log y - \int_y^x \frac{A(t)}{t}\,dt;$

(iii) $\displaystyle\sum_{r\le x} ra(r) = xA(x) - \int_1^x A(t)\,dt;$

(iv) $\displaystyle\sum_{r\le x} (x-r)a(r) = \int_1^x A(t)\,dt;$

(v) if $a(1) = 0$, then $\displaystyle\sum_{2\le r\le x} \frac{a(r)}{\log r} = \frac{A(x)}{\log x} + \int_2^x \frac{A(t)}{t(\log t)^2}\,dt.$

Example 2 To estimate $S(x) = \sum_{r\le x} a(r)/r$, given that $|A(x)| \le Cx$ for all x. By (i), we have

$$|S(x)| \le C + \int_1^x \frac{C}{t}\,dt = C(\log x + 1).$$

Continuous Abel summation has applications to infinite series exactly analogous to those of discrete Abel summation, as follows.

Proposition 1.3.8 *Suppose that f has a continuous derivative on $[1, \infty)$, and that $A(x)f(x) \to 0$ as $x \to \infty$. Then*

$$\sum_{r=1}^{\infty} a(r)f(r) = -\int_{1}^{\infty} A(t)f'(t)\,dt,$$

in the sense that if either side converges, then so does the other, to the same value. Further, we then have

$$\sum_{r>x} a(r)f(r) = -A(x)f(x) - \int_{x}^{\infty} A(t)f'(t)\,dt.$$

Proof The first statement follows from 1.3.6(i) on considering limits as x tends to ∞. The second statement follows similarly from 1.3.5, or by subtracting the identity in 1.3.6(i). □

Usually we will use this in the direction of deducing convergence of the series from that of the integral, as in the following corollary.

Corollary 1.3.9 *Suppose that:*

(i) $|A(t)| \le C$ *for all* t,

(ii) $f(t) \to 0$ *as* $t \to \infty$,

(iii) $\int_1^\infty |f'(t)|\,dt$ *is convergent (say to I).*

Then $\sum_{r=1}^{\infty} a(r)f(r)$ is convergent (say to S), and $|S| \le CI$.

Proof This follows from 1.3.8 and the general fact that if $\int_1^\infty |g(t)|dt$ converges (say to J), then so does $\int_1^\infty g(t)dt$, with value not greater than J in modulus. (Again, this remains valid in the complex case.) □

So we have, for example (given condition (i)):

$$\sum_{n=1}^{\infty} \frac{a(r)}{r} = \int_{1}^{\infty} \frac{A(t)}{t^2}\,dt.$$

Except when $f(t)$ is real and decreasing, the integral estimate for $|S|$ in 1.3.9 is usually more useful than the discrete one of 1.3.4, because it is easier to evaluate the integral than the discrete sum.

Finally, we apply Abel's formula to some specific choices of $a(r)$. As yet, our repertoire of arithmetic functions is very limited, but two cases are well worth mentioning. First, take $a(r) = 1$ for all r, so that the sum

$\sum_{r \le x} a(r)f(r)$ is simply $\sum_{r \le x} f(r)$. Then $A(x) = [x]$, and the two identities in 1.3.6 become

$$\sum_{r \le x} f(r) = [x]f(x) - \int_1^x [t]f'(t)\,dt,$$

$$\sum_{r \le x} [f(x) - f(r)] = \int_1^x [t]f'(t)\,dt.$$

Second, consider sums of the form $\sum_{p \in P[x]} f(p)$. Recall that $u_P(r)$ is 1 for r prime, 0 otherwise. With this notation, the stated sum can be rewritten $\sum_{r \le x} u_P(r)f(r)$. Clearly, $\sum_{r \le x} u_P(r) = \pi(x)$, so by 1.3.6(i),

$$\sum_{p \in P[x]} f(p) = \sum_{r \le x} u_P(r)f(r) = \pi(x)f(x) - \int_1^x \pi(t)f'(t)\,dt.$$

Since $\pi(t) = 0$ for $t < 2$, we can amend the interval of integration to $[2, x]$.

Example 3 To estimate $S(x) = \sum_{r \le x} \log(x/r)$.

By the second of the formulae just given for the case $a(r) = 1$,

$$S(x) = \sum_{r \le x} (\log x - \log r) = \int_1^x \frac{[t]}{t}\,dt.$$

To estimate this, write $t = [t] + \{t\}$, where $0 \le \{t\} < 1$. Then

$$S(x) = \int_1^x \frac{t - \{t\}}{t}\,dt = (x - 1) - q(x),$$

where

$$q(x) = \int_1^x \frac{\{t\}}{t}\,dt,$$

so that

$$0 \le q(x) \le \int_1^x \frac{1}{t}\,dt = \log x.$$

Exercises

1 Express $\sum_{p \in P[x]} (1/p)$ in terms of $\pi(x)$.

2 Let $S(x) = \sum_{r \le x} a(r) \log r$. Show that if $0 \le A(x) \le C$ for all $x \ge 1$, then $|S(x)| \le C \log x$ for all $x \ge 1$. *(Use either discrete or continuous Abel summation, or both!)*

3 Suppose that $|a(r)| \leq 1$ and $|A(n)| \leq C$ for all n. Using the appropriate integral expression, together with the fact that $|A(t)| \leq t$ for $1 \leq t \leq C$, show that

$$\left|\sum_{r=1}^{\infty} \frac{a(r)}{r}\right| \leq \log C + 1.$$

4 Assume (or prove) that if $B(n) \to B$ as $n \to \infty$, then

$$\frac{1}{n}[B(1) + \cdots + B(n)] \to B \quad \text{as } n \to \infty.$$

Now suppose that $\sum_{r=1}^{\infty} a(r)/r$ is convergent. Write $a(r)/r = b(r)$. Use discrete Abel summation to give an expression for $A(n)$ in terms of $B(n)$, and deduce that $A(n)/n \to 0$ as $n \to \infty$.

5 Suppose that $|A(t)| \leq C$ for all t, and that $f(t)$ is differentiable and decreasing and tends to 0 as $t \to \infty$. Prove that, for all $x \geq 1$,

$$\left|\sum_{r>x} a(r)f(r)\right| \leq 2Cf(x).$$

6 Prove that:

(a) if $|A(x)| \leq Cx/(\log x)^2$ for all $x \geq 2$, then $\sum_{r=1}^{\infty} a(r)/r$ is convergent;

(b) if $a(r) \geq 0$ for all r and $A(x) \geq Cx/\log x$ for all $x \geq 2$, then $\sum_{r=1}^{\infty} a(r)/r$ is divergent.

7 Let $b(r) = a(r)\log r$. Show that:

(a) if $|A(x)| \leq Cx$ for all $x \geq 1$, then $|B(x)| \leq Cx(\log x + 1)$ for all $x \geq 1$;

(b) if $a(1) = 0$ and $|B(x)| \leq Cx\log x$ for all $x \geq 2$, then $|A(x)| \leq C(x + \mathrm{li}(x))$ for all $x \geq 2$.

8 Prove that, for any real $x > 1$,

$$\sum_{r>x} \frac{1}{r^2} \leq \frac{1}{x} + \frac{1}{x^2}.$$

9 Show that for any real x, the partial sums $\sum_{r=1}^{n} \sin rx$ are bounded for all n (use an identity for $\cos(r - \frac{1}{2})x - \cos(r + \frac{1}{2})x$). Deduce that the series $\sum_{r=1}^{\infty} \frac{1}{r} \sin rx$ is convergent.

1.4 Estimation of sums by integrals; Euler's summation formula

Basic integral estimation

The use of an integral to estimate a discrete sum is a very useful technique, which we shall apply repeatedly. In its simplest form, the relationship is described in the following result.

Proposition 1.4.1 *Suppose that f is a decreasing function on $[m,n]$, where m,n are integers. Then*

$$f(m+1)+\cdots+f(n) \le \int_m^n f(t)\,dt \le f(m)+\cdots+f(n-1).$$

The opposite inequality applies if f is increasing.

Proof Suppose that $f(t)$ is decreasing. For $r-1 \le t \le r$, we have $f(r) \le f(t) \le f(r-1)$. So, by integration on this interval of length 1,

$$f(r) \le \int_{r-1}^{r} f(t)\,dt \le f(r-1).$$

By adding these inequalities for $r = m+1,\ldots,n$, we obtain the statement.

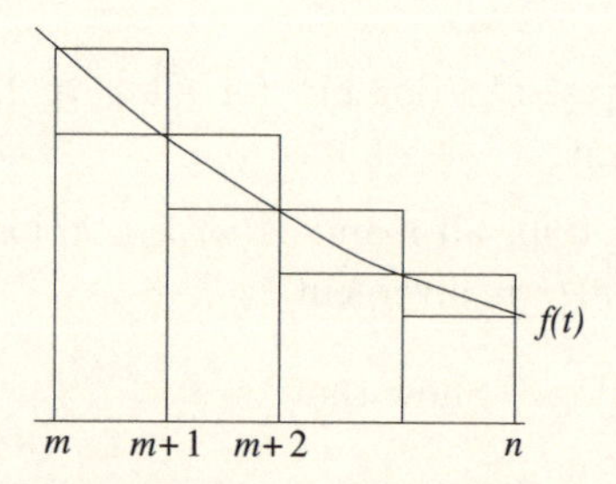

(This statement is well illustrated by the picture. The integral is the area under the curve, while the two sums are the total area of the small and big rectangles respectively.) □

For $x \ge 1$ (not necessarily an integer), write

$$S(x) = \sum_{1\le r\le x} f(r), \qquad I(x) = \int_1^x f(t)\,dt.$$

Taking $m = 1$ in 1.4.1, we obtain, in the case when $f(t)$ is decreasing,

$$f(2)+\cdots+f(n) \le \int_1^n f(t)\,dt \le f(1)+\cdots+f(n-1),$$

in other words,

$$S(n) - f(1) \leq I(n) \leq S(n-1), \tag{1.3}$$

or, equivalently,

$$I(n+1) \leq S(n) \leq I(n) + f(1), \tag{1.4}$$

with opposite inequalities when $f(t)$ is increasing. This shows how $I(n)$ can be used to give bounds for $S(n)$ when n is an integer. We shall often require estimates for $S(x)$ in terms of $I(x)$ when x is not an integer. Slightly different statements are appropriate for decreasing and increasing functions, as follows.

Proposition 1.4.2 *Suppose that $f(t)$ is non-negative and decreasing for all $t \geq 1$, and define $S(x)$, $I(x)$ as above. Then for all $x \geq 1$,*

$$I(x) \leq S(x) \leq I(x) + f(1).$$

Proof Suppose that $n \leq x < n+1$, where n is an integer. Then $S(x) = S(n)$ and $I(n) \leq I(x) \leq I(n+1)$. The statement now follows from (1.4). □

Example 1 Let $S(x) = \sum_{r \leq x} \frac{1}{r}$. Since

$$\int_1^x \frac{1}{t}\, dt = \log x,$$

we have $\log x \leq S(x) \leq \log x + 1$.

Proposition 1.4.3 *Suppose that $f(t)$ is non-negative and increasing for $t \geq 1$. Define $S(x)$, $I(x)$ as above. Then for all $x \geq 1$, we have $S(x) = I(x) + r(x)$, where $|r(x)| \leq f(x)$.*

Proof We now have $S(n) - f(n) \leq I(n) \leq S(n) - f(1)$ (the reverse of (1.3)). Let $n \leq x < n+1$. Then

$$S(x) = S(n) \leq I(n) + f(n) \leq I(x) + f(x).$$

Since $f(t)$ is increasing and $x - n < 1$, we have $\int_n^x f(t)\, dt \leq f(x)$, and hence

$$I(x) \leq I(n) + f(x) \leq S(n) - f(1) + f(x) \leq S(x) + f(x). \qquad \square$$

Note Of course, similar results apply if the sum and integral both start at 0 instead of 1 (in 1.4.2, $f(1)$ is then replaced by $f(0)$).

Example 2 Let $S(x) = \sum_{r \le x} \log r$. Then

$$I(x) = \int_1^x \log t \, dt = x \log x - x + 1,$$

and $|S(x) - I(x)| \le \log x$.

Even when we know $S(n)$ explicitly, as we do when $f(t) = t$, 1.4.3 gives a quick route to the estimation of $S(x)$ for non-integral x.

Returning to the case where $f(t)$ is decreasing, we derive the "integral test" for convergence of an infinite series.

Proposition 1.4.4 *Let $f(t)$ be decreasing and non-negative for all $t \ge 1$. Then the series $\sum_{r=1}^{\infty} f(r)$ is convergent if and only if the integral $\int_1^{\infty} f(t)dt$ is convergent. If the sum of the series is S and the value of the integral is I, then $I \le S \le I + f(1)$.*

Further, if we write

$$S^*(x) = \sum_{r > x} f(r), \qquad I^*(x) = \int_x^{\infty} f(t) \, dt,$$

then for integers n,

$$I^*(n+1) \le S^*(n) \le I^*(n),$$

and for general x,

$$I^*(x) - f(x) \le S^*(x) \le I^*(x) + f(x).$$

Proof Note that convergence of the series (or integral) means that $S(n)$ (or $I(n)$) tends to a limit as $n \to \infty$. Both $S(n)$ and $I(n)$ increase with n, so they tend to a limit if and only if they are bounded above. By (1.3),

$$S(n) - f(1) \le I(n) \le S(n-1).$$

If the integral converges to I, then $I(n) \le I$ for all n, hence $S(n) \le I + f(1)$. So $S(n) \to S$, where $S \le I + f(1)$. Conversely, if the series converges to S, then $I(n) \le S(n) \le S$ for all n, so $I(n) \to I$, where $I \le S$.

By 1.4.1 again, for integers n, p with $n < p$,

$$\int_{n+1}^{p+1} f(t) \, dt \le f(n+1) + \cdots + f(p) \le \int_n^p f(t) \, dt.$$

By letting $p \to \infty$, we obtain the stated inequalities for $S^*(n)$. Now let $n \le x < n+1$. Then $S^*(x) = S^*(n) \ge I^*(n+1)$. Now

$$I^*(n+1) = I^*(x) - \int_x^{n+1} f(t) \, dt,$$

and $\int_x^{n+1} f(t)\, dt \le f(x)$, since the length of the interval is less than 1. Hence $S^*(x) \ge I^*(x) - f(x)$. Also,

$$S^*(n) = S^*(n+1) + f(n+1) \le I^*(n+1) + f(n+1) \le I^*(x) + f(x). \quad \square$$

The best-known application of this result is convergence of the series $\sum_{n=1}^{\infty} 1/n^{\sigma}$ when $\sigma > 1$, which follows at once from the elementary fact that

$$\int_1^{\infty} \frac{1}{t^{\sigma}}\, dt = \frac{1}{\sigma - 1}.$$

This series defines the *Riemann zeta function*, which we will consider in more detail in section 1.7. Another example is the following.

Example 3 Consider the series

$$\sum_{n=2}^{\infty} \frac{1}{n \log n}, \qquad \sum_{n=2}^{\infty} \frac{1}{n(\log n)^2}.$$

Call the series S_1, S_2 respectively. The corresponding integrals are

$$\int_2^x \frac{1}{t \log t}\, dt = [\log \log t]_2^x = \log \log x - \log \log 2,$$

$$\int_2^x \frac{1}{t(\log t)^2}\, dt = \left[-\frac{1}{\log t} \right]_2^x = \frac{1}{\log 2} - \frac{1}{\log x}.$$

The first integral is divergent, while the second one converges to $1/\log 2$. Hence S_1 is divergent and S_2 is convergent.

We show next how example 1 combines with the expression given in 1.2.2 to give an estimation of $\sum_{r \le x} \tau(r)$ (a more accurate one will be given in section 2.5).

Proposition 1.4.5 *Let* $S_{\tau}(x) = \sum_{r \le x} \tau(r)$. *Then, for any* $x > 1$,

$$x(\log x - 1) \le S_{\tau}(x) \le x(\log x + 1).$$

Proof Recall from 1.2.2 that $S_{\tau}(x) = \sum_{r \le x} [x/r]$. Now $(x/r) - 1 \le [x/r] \le (x/r)$, so

$$x \sum_{r \le x} \frac{1}{r} - x \le S_{\tau}(x) \le x \sum_{r \le x} \frac{1}{r}.$$

By example 1, the sum $\sum_{r \le x} \frac{1}{r}$ lies between $\log x$ and $\log x + 1$. The statement follows. $\square$

Euler's summation formula

Euler's summation formula gives an actual expression for the difference between a sum and the corresponding integral. We shall give three variants of the formula. The most basic one is as follows.

Proposition 1.4.6 (Euler's summation formula, version 1) *Let m,n be integers and let f be a differentiable function on the interval $[m, n]$ (with real or complex values). Then*

$$\sum_{r=m+1}^{n} f(r) - \int_{m}^{n} f(t)\,dt = \int_{m}^{n} (t - [t]) f'(t)\,dt.$$

Proof For $r - 1 \leq t < r$, we have $[t] = r - 1$. Integration by parts gives

$$\begin{aligned}\int_{r-1}^{r} (t - r + 1) f'(t)\,dt &= [(t - r + 1) f(t)]_{r-1}^{r} - \int_{r-1}^{r} f(t)\,dt \\ &= f(r) - \int_{r-1}^{r} f(t)\,dt.\end{aligned}$$

This equates to $\int_{r-1}^{r} (t - [t]) f'(t)\,dt$, since the new value of $[t]$ at the single point r makes no difference. Add for $r = m + 1, \ldots, n$ to obtain the statement. □

The second version is symmetrical, taking half the values at the end points. At the cost of a slightly more complicated expression, it describes the difference between the integral and the "trapezium rule" approximation (which is normally much closer) instead of the rectangular one.

Proposition 1.4.7 (version 2) *Under the same conditions, we have*

$$\tfrac{1}{2} f(m) + \sum_{r=m+1}^{n-1} f(r) + \tfrac{1}{2} f(n) - \int_{m}^{n} f(t)\,dt = \int_{m}^{n} (t - [t] - \tfrac{1}{2}) f'(t)\,dt.$$

Proof This follows easily from 1.4.6, or else it can be proved in the same way, modified by

$$[(t - r + \tfrac{1}{2}) f(t)]_{r-1}^{r} = \tfrac{1}{2}[f(r - 1) + f(r)].$$ □

Note Denote the quantities in the statements of 1.4.6 and 1.4.7 by $\Delta(m, n)$ and $\Delta'(m, n)$ respectively. Since $0 \leq t - [t] < 1$ and $|t - [t] - \frac{1}{2}| \leq \frac{1}{2}$ for all t, we have

$$|\Delta(m, n)| \leq \int_{m}^{n} |f'(t)|\,dt, \qquad |\Delta'(m, n)| \leq \tfrac{1}{2} \int_{m}^{n} |f'(t)|\,dt.$$

(We are using the elementary fact that $|\int_a^b g(t)\,dt| \le \int_a^b |g(t)|\,dt$; for a justification of this in the case when $g(t)$ is complex, see appendix A.) We shall see below that a much stronger estimate actually applies to $\Delta'(m,n)$. It also follows from 1.4.5 that if $f(t)$ is increasing, then

$$0 \le \Delta(m,n) \le \int_m^n f'(t)\,dt = f(n) - f(m),$$

which gives a second proof of the increasing case in 1.4.1.

The third version replaces n by a non-integer x, at the cost of an extra term. (To do the same for m, one can then just take the difference between two such expressions.)

Proposition 1.4.8 (version 3) *Let m be an integer and x real. Let f be differentiable on $[m,x]$. Then*

$$\sum_{m<r\le x} f(r) - \int_m^x f(t)\,dt = \int_m^x (t-[t])f'(t)\,dt - (x-[x])f(x).$$

Proof Let $[x]=n$. To the sum considered in 1.4.6 we now add

$$\begin{aligned}\int_n^x (t-n)f'(t)\,dt &= [(t-n)f(t)]_n^x - \int_n^x f(t)\,dt \\ &= (x-n)f(x) - \int_n^x f(t)\,dt.\end{aligned}$$ □

Note This version of Euler's summation formula can be derived from Abel's summation formula for $\sum_{r\le x} 1.f(r)$ together with integration by parts of $\int_1^x 1.f(t)\,dt$. However, this is no simpler than the direct proof just given.

In the case when the series and integral are convergent, we can take the limit as $N\to\infty$ in 1.4.6 and 1.4.7, obtaining:

Proposition 1.4.9 *Suppose that f is differentiable on $[1,\infty)$ and that both $\sum_{r=1}^\infty f(r)$ and $\int_1^\infty f(t)\,dt$ are convergent. Then*

$$\begin{aligned}\sum_{r=1}^\infty f(r) - \int_1^\infty f(t)\,dt &= f(1) + \int_1^\infty (t-[t])f'(t)\,dt \\ &= \tfrac{1}{2}f(1) + \int_1^\infty (t-[t]-\tfrac{1}{2})f'(t)\,dt.\end{aligned}$$

Proof For the first expression, take $m=1$ in 1.4.6 and let $n\to\infty$. Note that the sum on the left-hand side starts with $f(2)$, so we must add $f(1)$.

The second expression follows similarly from 1.4.7, or can be deduced easily from the first expression. □

We now show that for decreasing, real-valued functions, even when the series and integral diverge, the difference $S(x) - I(x)$ tends to a limit. This enables us to proceed to a further degree of accuracy in the estimation of the sum $S(x)$.

Proposition 1.4.10 *Suppose that $f(t)$ is decreasing and tends to 0 as $t \to \infty$. Define $S(x)$ and $I(x)$ as above. Then $S(x) - I(x) \to L$ as $x \to \infty$, where*

$$L = f(1) + \int_1^\infty (t - [t])f'(t)\,dt.$$

We have $0 \le L \le f(1)$ and for all $x \ge 1$,

$$S(x) = I(x) + L + q(x),$$

where $|q(x)| \le f(x)$. If x is an integer n, then $0 \le q(x) \le f(x)$.

Proof Take $m = 1$ in 1.4.8 and add $f(1)$ to both sides. We obtain

$$S(x) - I(x) = f(1) + \int_1^x (t - [t])f'(t)\,dt - (x - [x])f(x).$$

Now

$$\int_a^\infty f'(t)\,dt = \lim_{R\to\infty} \int_a^R f'(t)\,dt = \lim_{R\to\infty} [f(R) - f(a)] = -f(a)$$

(note that $f'(t)$ is negative!). Since $0 \le t - [t] < 1$ for all t, the integral $\int_1^\infty (t - [t])f'(t)\,dt$ converges, with value between $-f(1)$ and 0. Hence $S(x) - I(x)$ tends to the stated limit L, and $0 \le L \le f(1)$. Further,

$$S(x) - I(x) = L - \int_x^\infty (t - [t])f'(t)\,dt - (x - [x])f(x).$$

Write this as $L + J(x) - F(x)$. Then

$$0 \le J(x) \le -\int_x^\infty f'(t)\,dt = f(x).$$

Also, $0 \le F(x) \le f(x)$, so $|J(x) - F(x)| \le f(x)$ (and if x is an integer, then $F(x) = 0$). □

The most important particular case of this situation is the following, obtained by taking $f(t) = 1/t$.

Proposition 1.4.11 *The expression* $\sum_{r=1}^{n} \frac{1}{r} - \log n$ *tends to the limit* γ *as* $n \to \infty$, *where*

$$\gamma = 1 - \int_1^\infty \frac{t - [t]}{t^2}\, dt$$

(γ is called Euler's constant). Further, $0 < \gamma < 1$ *and*

$$\sum_{1 \le r \le x} \frac{1}{r} = \log x + \gamma + q(x),$$

where $|q(x)| \le 1/x$. □

As we shall see, the constant γ arises constantly in number-theoretic estimations. Its value is actually $0.577216\ldots$.

Closer estimation for version 2

We now show the true extent of the closer approximation given by the symmetric form of Euler's formula. Among other things, this will give us an expression that converges much more rapidly to γ. Apart from this, the closer estimate will only be used for some results on the zeta function that are not essential for the prime number theorem, so although this topic (in the author's opinion!) is very interesting, the reader is at liberty to omit it.

Proposition 1.4.12 *Suppose that f is twice differentiable on* $[m, n]$. *Denote the quantity in 1.4.7 by* $\Delta'(m, n)$. *Then*

$$|\Delta'(m, n)| \le \tfrac{1}{8} \int_m^n |f''(t)|\, dt.$$

If f is real-valued and $f''(t) \ge 0$ *on the interval, then*

$$0 \le \Delta'(m, n) \le \tfrac{1}{8}[f'(n) - f'(m)].$$

Proof Where we previously integrated by parts, we now do so the other way round! Writing $n - \frac{1}{2} = c$, we have

$$\begin{aligned}
\int_{r-1}^{r} (t - c) f'(t)\, dt &= [\tfrac{1}{2}(t - c)^2 f'(t)]_{r-1}^{r} - \tfrac{1}{2} \int_{r-1}^{r} (t - c)^2 f''(t)\, dt \\
&= \tfrac{1}{8}[f'(r) - f'(r - 1)] - \tfrac{1}{2} \int_{r-1}^{r} (t - c)^2 f''(t)\, dt \\
&= \int_{r-1}^{r} [\tfrac{1}{8} - \tfrac{1}{2}(t - c)^2] f''(t)\, dt.
\end{aligned}$$

Now $0 \le \frac{1}{2}(t-c)^2 \le \frac{1}{8}$ for t in $[r-1, r]$. Both the stated estimations follow. □

If $f''(t) \ge 0$ on the interval, then the function f is "convex" there, and the inequality $\Delta'(m,n) \ge 0$ represents the geometrically obvious fact that, for such a function, the trapezium estimate is not less than the integral.

We now formulate the more accurate variant of 1.4.10 corresponding to this approach.

Proposition 1.4.13 *With the assumptions and notation of 1.4.10, suppose further that* $f''(t) \ge 0$ *for all* $t \ge 1$. *For integers n, write*

$$\tilde{S}(n) = \sum_{r=1}^{n-1} f(r) + \tfrac{1}{2} f(n)$$

and $B(x) = t - [t] - \frac{1}{2}$. *Then*

$$L = \tfrac{1}{2} f(1) + \int_1^\infty B(t) f'(t)\, dt,$$

and

$$\tfrac{1}{2} f(1) \le L \le \tfrac{1}{2} f(1) - \tfrac{1}{8} f'(1).$$

Also, for integers $n \ge 1$,

$$\tilde{S}(n) = I(n) + L - \tilde{q}(n),$$

where $0 \le \tilde{q}(n) \le -\frac{1}{8} f'(n)$.

Proof From 1.4.7, we have

$$\tilde{S}(n) - I(n) = \tfrac{1}{2} f(1) + \int_1^n B(t) f'(t)\, dt.$$

Since $f(n) \to 0$ as $n \to \infty$, we have

$$L = \lim_{n\to\infty} [\tilde{S}(n) - I(n)] = \tfrac{1}{2} f(1) + \int_1^\infty B(t) f'(t)\, dt,$$

so by subtraction

$$\tilde{S}(n) - I(n) - L = -\int_n^\infty B(t) f'(t)\, dt.$$

If $f''(t) \ge 0$, then 1.4.12 (with n tending to ∞ and m replaced by n) shows that the value of this last integral is between 0 and $-\frac{1}{8} f'(n)$. In the same way, the expression for L itself shows that L is between $\frac{1}{2} f(1)$ and $\frac{1}{2} f(1) - \frac{1}{8} f'(1)$. □

Applied with $f(t) = 1/t$, this gives $\frac{1}{2} \leq \gamma \leq \frac{5}{8}$, and further,

$$\sum_{r=1}^{n-1} \frac{1}{r} + \frac{1}{2n} = \log n + \gamma - \tilde{q}(n),$$

where $0 \leq \tilde{q}(n) \leq 1/(8n^2)$. So $\log n + 1/(2n) + \gamma$ is a much better approximation to $\sum_{r=1}^{n} \frac{1}{r}$ than $\log n + \gamma$. Equally, for the purpose of evaluating γ, there is a great gain in accuracy in using the expression in which the term $1/n$ is replaced by $1/(2n)$. For example, if this is done with $n = 10$, we will obtain an approximation to γ with error less than $1/800$. An error of $1/20$ would immediately be added if the tenth term were not halved.

Exercises

1 Give upper and lower estimates for $\sum_{r \leq x} 1/r^{1/2}$.

2 Give estimates for $\sum_{r \leq x} r$ and $\sum_{r \leq x} r^{1/2}$, using integrals on $[0, x]$. Compare the first estimate with the known sum $\sum_{r \leq n} r$, where $n \leq x < n + 1$.

3 (*Alternative proof of 1.4.10 without Euler's formula*) Suppose that $f(t)$ is decreasing and tends to 0 as $t \to \infty$. Show that (i) $S(n) - I(n)$ is decreasing, (ii) $S(n-1) - I(n)$ is increasing, and hence that both these expressions tend to a limit L as $n \to \infty$, where $0 \leq L \leq f(1)$. Show further that $L \leq S(n) - I(n) \leq L + f(n)$ for integers $n \geq 1$.

4 Use integration by parts to find an indefinite integral of $(\log x)^2$. Now prove that

$$\sum_{r \leq x} (\log x - \log r)^2 = 2x + q(x),$$

where $|q(x)| \leq 4(\log x)^2 + 2\log x + 2$.

5 Let $S(x) = \sum_{r \leq x} 1/r^{1/2}$. Show that $S(x) - 2x^{1/2} + 2$ tends to a limit L as $x \to \infty$. By taking $x = 4$ and applying 1.4.13, show that L lies between 0.534 and 0.543.

6 In 1.4.12, suppose that f is real-valued and that $m_r \leq f''(t) \leq M_r$ on $[r-1, r]$, where $m_r \geq 0$. By evaluating the integral obtained in the proof, show that

$$\frac{1}{12} \sum_{r=m+1}^{n} m_r \leq \Delta'(m, n) \leq \frac{1}{12} \sum_{r=m+1}^{n} M_r.$$

1.5 The function li(x)

Since it is being proposed as the answer to our fundamental question, the function li(x) merits some attention. Recall the definition: for $x \geq 2$,

$$\mathrm{li}(x) = \int_2^x \frac{1}{\log t}\, dt.$$

For $x < 2$, we define li(x) to be 0. There is no "elementary" formula for the integral above, which is why the notation li(x) (which is standard) is introduced. Note that by the fundamental theorem of calculus, the derivative of li(x) (for $x \geq 2$) is $1/\log x$.

There is a closely related discrete sum, denoted by ls(x) (in which ls stands for "logarithmic sum"):

$$\mathrm{ls}(x) = \sum_{2 \leq n \leq x} \frac{1}{\log n}.$$

To compare the two quantities, we apply 1.4.2. In the notation used there, $\mathrm{li}(x) = I(x)$ and $\mathrm{ls}(x) = S(x)$, with $f(t) = 1/(\log t)$ and the sum starting at 2. Hence we have:

Proposition 1.5.1 . *For all $x \geq 2$, we have*

$$\mathrm{li}(x) \leq \mathrm{ls}(x) \leq \mathrm{li}(x) + \frac{1}{\log 2}. \qquad \square$$

We now investigate the relationship between li(x) and expressions like $x/(\log x)$. First, since $(1/\log t) \geq (1/\log x)$ for $2 \leq t \leq x$, it is clear that

$$\mathrm{li}(x) \geq \frac{x-2}{\log x}$$

for all $x \geq 2$. We will show that $\mathrm{li}(x) \sim (x/\log x)$ as $x \to \infty$, while at the same time obtaining an asymptotic estimate of the difference. Integration by parts, with 1 as one factor, gives

$$\mathrm{li}(x) = \frac{x}{\log x} - \alpha + \int_2^x \frac{1}{(\log t)^2}\, dt, \qquad (1.5)$$

where $\alpha = 2/(\log 2)$. A further integration by parts would give an expression involving the integral of $1/(\log t)^3$. To explore this, write, for each $n \geq 1$,

$$I_n(x) = \int_e^x \frac{1}{(\log t)^n}\, dt.$$

We start the integration at e to avoid the appearance of tiresome constants;

it has the effect that $I_1(x)$ differs from $\mathrm{li}(x)$ by a small constant (less than 1). Integration by parts gives

$$\begin{aligned} I_n(x) &= \left[\frac{t}{(\log t)^n}\right]_e^x + \int_e^x t\frac{n}{t(\log t)^{n+1}}\,dt \\ &= \frac{x}{(\log x)^n} - e + nI_{n+1}(x). \end{aligned} \tag{1.6}$$

Recall that for any $k \geq 1$,

$$\frac{(\log x)^k}{x} \to 0 \qquad \text{as } x \to \infty$$

(we shall use this well-known fact frequently, both here and in later sections).

Proposition 1.5.2 *We have* $I_n(x) \sim \dfrac{x}{(\log x)^n}$ *as* $x \to \infty$.

Proof By (1.6), the statement is equivalent to saying that

$$I_{n+1}(x)\frac{(\log x)^n}{x} \to 0 \qquad \text{as } x \to \infty.$$

Divide the interval into two pieces at $x^{1/2}$. Since $\log t \geq 1$ for $e \leq t \leq x^{1/2}$ and $\log t \geq \frac{1}{2}\log x$ for $x^{1/2} \leq t \leq x$,

$$\begin{aligned} I_{n+1}(x) &= \int_e^{x^{1/2}} \frac{1}{(\log t)^{n+1}}\,dt + \int_{x^{1/2}}^x \frac{1}{(\log t)^{n+1}}\,dt \\ &\leq x^{1/2} + x\left(\frac{2}{\log x}\right)^{n+1}, \end{aligned}$$

in which we have used overestimates $x^{1/2}$ and x for the lengths of the two intervals. So we have

$$I_{n+1}(x)\frac{(\log x)^n}{x} \leq \frac{(\log x)^n}{x^{1/2}} + \frac{2^{n+1}}{\log x},$$

which tends to 0 as $x \to \infty$ (as required). □

Proposition 1.5.3 *We have* $\mathrm{li}(x) \sim x/\log x$ *as* $x \to \infty$. *More precisely,*

$$\mathrm{li}(x) = \frac{x}{\log x} + r(x),$$

where

$$r(x) \sim \frac{x}{(\log x)^2} \qquad \text{as } x \to \infty.$$

Proof It is enough to prove the statements for $I_1(x)$ instead of $\mathrm{li}(x)$. The first statement is 1.5.2 with $n = 1$. By (1.6),

$$I_1(x) = \frac{x}{\log x} + I_2(x) - e. \tag{1.7}$$

By 1.5.2 again, $I_2(x) - e \sim x/(\log x)^2$ as $x \to \infty$. □

Proposition 1.5.4 *We have*

$$\mathrm{li}(x) = \frac{x}{\log x - 1} + q(x),$$

where

$$q(x) \sim \frac{x}{(\log x)^3} \quad \text{as } x \to \infty.$$

Proof Again replace $\mathrm{li}(x)$ by $I_1(x)$. Using (1.6) to substitute for $I_2(x)$ in (1.7), we obtain

$$I_1(x) = \frac{x}{\log x} + \frac{x}{(\log x)^2} - 2e + 2I_3(x).$$

Now

$$\frac{1}{y} + \frac{1}{y^2} - \frac{1}{y-1} = \frac{(y+1)(y-1) - y^2}{(y-1)y^2} = -\frac{1}{(y-1)y^2}.$$

Substituting this with $y = \log x$ gives

$$q(x) = I_1(x) - \frac{x}{\log x - 1} = -\frac{x}{(\log x - 1)(\log x)^2} - 2e + 2I_3(x),$$

so by 1.5.3 again,

$$q(x)\frac{(\log x)^3}{x} \to -1 + 0 + 2 = 1 \qquad \text{as } x \to \infty. \qquad \square$$

The last two results show that, as an approximation to $\mathrm{li}(x)$, the function $x/(\log x - 1)$ is better than $x/\log x$ by a factor of $\log x$. Presupposing that $\mathrm{li}(x)$ itself is a good approximation to $\pi(x)$, this explains the relative accuracies seen in the table in section 1.1.

Clearly, 1.5.4 also shows that $\mathrm{li}(x) > x/(\log x - 1)$ for all large enough x.

It follows from 1.5.3 that, for some constant C, we have $\mathrm{li}(x) \le Cx/\log x$ for all $x \ge 2$. For certain purposes, it will be useful to have a definite estimate of C in the form provided by the next result.

Proposition 1.5.5 *Write $\alpha = 2/(\log 2)$. Then, for all $x \ge 2$,*

$$\frac{x}{\log x} \le \mathrm{li}(x) + \alpha \le C\frac{x}{\log x},$$

where $C \le 3\frac{1}{10}$.

Proof The left-hand inequality follows at once from the elementary inequality $\mathrm{li}(x) \geq (x-2)/(\log x)$. For the right-hand inequality, again split the interval $[2, x]$ at $x^{1/2}$. Since $\log t \geq \log 2$ on $[2, x^{1/2}]$ and $\log t \geq \frac{1}{2}\log x$ on $[x^{1/2}, x]$, we deduce

$$\begin{aligned} \mathrm{li}(x) &= \int_2^{x^{1/2}} \frac{1}{\log t}\, dt + \int_{x^{1/2}}^{x} \frac{1}{\log t}\, dt \\ &\leq \frac{x^{1/2}-2}{\log 2} + \frac{2x}{\log x}. \end{aligned}$$

By differentiation, one finds that the greatest value of $(\log x)/x^{1/2}$ is $2/e$ (occurring when $x = e^2$). Hence $x^{1/2} \leq (2/e)(x/\log x)$. The statement follows, since $2/(e\log 2) < 1\frac{1}{10}$. □

With more effort, one can show that C is actually less than 2.

Computation of $\mathrm{li}(x)$. This is best done by methods of numerical integration such as Simpson's rule, rather than by continuing the above process of integration by parts. It should be remembered that numerical approximation has to be used to find values of all the familiar functions that are given by series or integrals, such as e^x and $\log x$. In fact, because $1/\log t$ varies very slowly, Simpson's rule gives excellent accuracy for its integral on intervals $[a, b]$ where b is not greater than $2a$. Recall that Simpson's rule approximates to $\int_a^b f$ by

$$\tfrac{1}{6}[f(a) + 4f(c) + f(b)],$$

where $c = \frac{1}{2}(a+b)$.

Example Let $a = 500{,}000$ and $b = 1{,}000{,}000$. The slow variation of $1/\log t$ is illustrated by

$$\frac{1}{\log a} \approx 0.076206, \qquad \frac{1}{\log b} \approx 0.072382.$$

The estimate for $\int_a^b (1/\log t)\, dt$ given by Simpson's rule is 37,023, which differs from the exact value by less than 2. As shown by the table in section 1.1, the number of primes in this interval is 36,960.

The tedious part of the calculation is getting started. Given (for free) the value $\mathrm{li}(1000) = 176.6$, you may care to do some calculations for yourself. Of course, tabulated values of $\mathrm{li}(x)$ have been published.

By contrast, direct calculation of $\operatorname{ls}(x)$ for large x would be extremely laborious! The only sensible way to estimate it is by derivation from $\operatorname{li}(x)$.

Note on the definition of $\operatorname{li}(x)$. For more refined purposes, it is customary to define $\operatorname{li}(x)$ to be the "Cauchy principal value" of $\int_0^x (1/\log t)\,dt$, that is (for $x > 1$):

$$\lim_{\delta \to 0+} \left(\int_0^{1-\delta} + \int_{1+\delta}^{x} \right) \frac{1}{\log t}\,dt.$$

This simply adds a small constant (approximately 1.04) to our definition of $\operatorname{li}(x)$, so it makes no difference to statements of the form $\pi(x) \sim \operatorname{li}(x)$ (though requiring a slight adjustment to results like 1.5.1). For our purposes, it is an unnecessary complication.

Exercises

1 Show directly from the defining series that $\operatorname{ls}(n) > \dfrac{n}{\log n}$ for $n \geq 4$.

2 Integrate by parts to show that

$$\int_2^x \frac{\operatorname{li}(t)}{t}\,dt = \operatorname{li}(x)\log x - (x-2).$$

3 Show that, for $a > 1$,

$$\int_2^\infty \frac{1}{t^a \log t}\,dt = a \int_2^\infty \frac{\operatorname{li}(t)}{t^{a+1}}\,dt.$$

4 By repeated application of formula (1.6), show that, for each n,

$$\operatorname{li}(x) = \frac{x}{\log x} + \frac{x}{(\log x)^2} + \cdots + (n-1)!\frac{x}{(\log x)^n} + r_{n+1}(x),$$

where

$$r_{n+1}(x) \sim n!\frac{x}{(\log x)^{n+1}} \quad \text{as } x \to \infty.$$

5 Show that, for all $a > 0$,

$$\operatorname{li}(x^a) = \int_{2^{1/a}}^x \frac{u^{a-1}}{\log u}\,du.$$

6 Let $0 < \alpha < 1$, and for $n \geq 1$ let

$$J_n(x) = \int_2^x \frac{1}{t^\alpha (\log t)^n}\,dt.$$

Integrate by parts to express $J_n(x)$ in terms of $J_{n+1}(x)$. Choose $\delta > 0$

such that $\alpha + \delta < 1$. Determine the number c_n such that $t^\delta/(\log t)^n$ is increasing for all $t \geq c_n$. By splitting the interval $[2, x]$ into $[2, c_n]$ and $[c_n, x]$, and writing $1/t^\alpha$ as $t^\delta/t^{\alpha+\delta}$, show that

$$J_n(x)\frac{(\log x)^{n-1}}{x^{1-\alpha}} \to 0 \qquad \text{as } x \to \infty,$$

and hence that

$$J_n(x) \sim \frac{x^{1-\alpha}}{(1-\alpha)(\log x)^n} \qquad \text{as } x \to \infty.$$

(This gives an alternative method for 1.5.2.)

1.6 Chebyshev's theta function

The first real progress towards the prime number theorem was achieved by Chebyshev in 1850. In a flash of inspiration, he saw that, instead of simply counting primes, it might be easier to count them with "weights". In other words, take a function f and consider sums like $\sum_{p \in P[x]} f(p)$, where again $P[x]$ denotes the set of primes not greater than x. As we have seen, Abel summation can then be used to relate such quantities to $\pi(x)$ itself.

In particular, Chebyshev considered the function $\theta(x)$ defined in this way, with $f(x)$ chosen to be $\log x$:

$$\theta(x) = \sum_{p \in P[x]} \log p.$$

The idea is very roughly that the weight $\log p$ will compensate for the decreasing density of primes, which we conjecture to be (in some sense) $1/\log n$ around the integer n. So we expect the growth of $\theta(x)$ to be linear.

The notation $\theta(x)$ is now standard. Another way to describe it is: if $p_1, p_2, \ldots, p_n$ are the primes not greater than x, then

$$\theta(x) = \log p_1 + \cdots + \log p_n = \log(p_1 p_2 \ldots p_n).$$

So by considering $\theta(x)$ we are effectively considering the quite natural quantity $p_1 p_2 \ldots p_n$. An obvious inequality is

Proposition 1.6.1 *For all $x > 0$, we have $\theta(x) \leq \pi(x)\log x$.*

Proof With the above notation, $n = \pi(x)$ and $\log p_j \leq \log x$ for each $j \leq n$. □

The relationship is expressed more exactly by Abel's summation formula. Recall that $u_P(n)$ is 1 if n is prime and 0 otherwise. By 1.3.6, we have

$$\theta(x) = \sum_{n \le x} u_P(n) \log n = \pi(x) \log x - \int_2^x \frac{\pi(t)}{t} \, dt.$$

If $\pi(x)$ really is roughly $x/\log x$, then $\theta(x)$ will be roughly x (the integral term will be roughly $\mathrm{li}(x)$, hence small compared to x).

Conversely, we can express $\pi(x)$ in terms of $\theta(x)$. Clearly, $\theta(x) = \sum_{n \le x} b(n)$, where

$$b(n) = \begin{cases} \log n & \text{if } n \text{ is prime,} \\ 0 & \text{otherwise.} \end{cases}$$

Since $b(n)/\log n$ is 1 for n prime, we have

$$\pi(x) = \sum_{2 \le n \le x} \frac{b(n)}{\log n}.$$

Since $b(1) = 0$, version 1.3.7 of Abel's summation formula gives

$$\pi(x) = \frac{\theta(x)}{\log x} + \int_2^x \frac{\theta(t)}{t(\log t)^2} \, dt \tag{1.8}$$

for all $x \ge 2$. Using this expression, we now show more precisely how an estimation for $\theta(x)$ will lead to one for $\pi(x)$. This is one of the fundamental steps on our journey to the prime number theorem.

Theorem 1.6.2 *Suppose that for some constants c_0, C_0 we have*

$$c_0 x \le \theta(x) \le C_0 x$$

for all $x \ge 2$. Then

$$c_0 \left[\mathrm{li}(x) + \alpha\right] \le \pi(x) \le C_0 \left[\mathrm{li}(x) + \alpha\right]$$

for all $x \ge 2$, where $\alpha = 2/\log 2$. Further, suppose that for some constants c, C and some x_0, we have $cx \le \theta(x) \le Cx$ for all $x \ge x_0$. Let $\varepsilon > 0$. Then there exists x_1 such that

$$(c - \varepsilon)\mathrm{li}(x) \le \pi(x) \le (C + \varepsilon)\mathrm{li}(x)$$

for all $x \ge x_1$.

Proof Integration by parts, with 1 as one factor, gives

$$\mathrm{li}(x) = \int_2^x \frac{1}{\log t} \, dt = \frac{x}{\log x} - \alpha + \int_2^x \frac{1}{(\log t)^2} \, dt. \tag{1.9}$$

The first statement now follows at once by comparing (1.8) and (1.9). For the second statement, take $x > x_0$. Then

$$\begin{aligned}\pi(x) &= \frac{\theta(x)}{\log x} + \int_2^{x_0} \frac{\theta(t)}{t(\log t)^2}\,dt + \int_{x_0}^{x} \frac{\theta(t)}{t(\log t)^2}\,dt \\ &= \frac{\theta(x)}{\log x} + C\int_2^{x_0} \frac{1}{(\log t)^2}\,dt + K + \int_{x_0}^{x} \frac{\theta(t)}{t(\log t)^2}\,dt,\end{aligned}$$

where

$$K = \int_2^{x_0} \left(\frac{\theta(t)}{t(\log t)^2} - \frac{C}{(\log t)^2} \right) dt.$$

Hence

$$\begin{aligned}\pi(x) &\le \frac{Cx}{\log x} + C\int_2^{x} \frac{1}{(\log t)^2}\,dt + K \\ &= C\,(\mathrm{li}(x) + \alpha) + K \\ &= C\,\mathrm{li}(x) + K',\end{aligned}$$

where $K' = K + \alpha C$. Since $\mathrm{li}(x) \to \infty$ as $x \to \infty$, there exists x_1 such that, for $x \ge x_1$, we have $K' \le \varepsilon\,\mathrm{li}(x)$, so that $\pi(x) \le (C+\varepsilon)\mathrm{li}(x)$. The left-hand inequality is proved in exactly the same way. □

Note A similar proof using discrete Abel summation delivers a variant of the theorem in terms of $\mathrm{ls}(x)$ instead of $\mathrm{li}(x)$.

A numerical example. Some actual numbers will help to illustrate how $\theta(x)$ proceeds in "jerks", and the limits to how well it can be expected to approximate to x. Computation gives $\theta(181) \approx 167.47$. There are no primes between 181 and 191, so $\theta(x)$ remains constant at this value for $181 \le x < 191$. The numbers 191, 193, 197, 199 are all prime, each adding slightly more than 5 to $\theta(x)$, with the result that $\theta(199) \approx 188.56$. This value then persists until we reach 211, the next prime.

Since the jumps $\log p$ in the value of $\theta(x)$ become increasingly large, it is clear that the difference $\theta(x) - x$ is unbounded, even if (as we hope) it is ultimately small compared to x.

To prove the prime number theorem, we will have to show that $\theta(x)$ satisfies the hypothesis of 1.6.2 with $c = 1 - \varepsilon$ and $C = 1 + \varepsilon$ (for any given $\varepsilon > 0$). Apart from diversions, this task will occupy most of the next two chapters. However, one can show by much more direct methods that the stated inequalities hold for *some* c and C. This is what Chebyshev did. Our first proof of the prime number theorem does not actually need these results, while our second and third proofs require the upper estimate but not the lower one. However, the estimates (particularly the upper one) have numerous other applications. Also, their proofs are both ingenious and instructive. Here we prove the upper estimate, leaving the lower one until later (section 2.4), when we will have the machinery to prove it much more neatly.

The proof is rather more number-theoretic in nature than anything we have encountered yet. In particular, we need to recall two statements from elementary number theory.

(NT1) *If p is prime and $p|ab$, then $p|a$ or $p|b$.*

(NT2) *If $p_1|a$, $p_2|a$ and the greatest common divisor of p_1 and p_2 is 1 (in particular, if p_1, p_2 are distinct primes), then $p_1p_2|a$.*

These statements are easy consequences of unique prime factorization, but in the usual approach they are actually used in proving it.

Note that by (NT1), if $p|n!$, then $p|k$ for some $k \leq n$, so certainly $p \leq n$.

Theorem 1.6.3 *For all $x > 1$, we have*

$$\theta(x) \leq (\log 4)x,$$

$$\pi(x) \leq (\log 4)\operatorname{li}(x) + 4.$$

Note In fact, $\log 4 \approx 1.3863$: our constant C is not much bigger than 1.

Proof It is sufficient to consider integers n. Fix n, and let

$$N = \binom{2n+1}{n} = \frac{(2n+1)(2n)\dots(n+2)}{n!}.$$

Now $\binom{2n+1}{n} = \binom{2n+1}{n+1}$, and these are two terms from the binomial expansion of $2^{2n+1} = (1+1)^{2n+1}$. Hence $N < 2^{2n} = 4^n$. (Of course, this is a very weak estimation of N, but it is all we need for the present proof; for an alternative proof, see exercise 3.)

Let $p_{k+1}, \dots, p_m$ be the primes p such that $n+2 \leq p \leq 2n+1$, so that

$$\sum_{j=k+1}^{m} \log p_j = \theta(2n+1) - \theta(n+1).$$

By (NT1), these primes do not divide into $n!$. But they do divide into $(2n+1)\dots(n+2) = n!N$, so by (NT1) again, they divide into N. By (NT2), their product $p_{k+1}\dots p_m$ divides into N, so is not greater than N. Therefore

$$\theta(2n+1) - \theta(n+1) = \log(p_{k+1}\cdots p_m) \leq \log N \leq n \log 4. \tag{1.10}$$

It is easily verified that $\theta(n) \leq n \log 4$ for $n = 2, 3, 4$. Suppose for induction that $\theta(k) \leq k \log 4$ for all $k \leq 2n$, where $n > 1$. Then, in particular, $\theta(n+1) \leq (n+1)\log 4$, so by (1.10) we have $\theta(2n+1) \leq (2n+1)\log 4$. Also, $\theta(2n+2) = \theta(2n+1)$, since $2n+2$ is not prime. Hence the required inequality holds for all $k \leq 2n+2$, completing the induction step.

The statement for $\pi(x)$ follows, by 1.6.2, since $\alpha \log 4 = 4$. □

Corollary 1.6.4 *There is a constant $C_1 \leq 3\frac{1}{10} \log 4$ such that*

$$\pi(x) \leq C_1 \frac{x}{\log x} \qquad \text{for all } x \geq 2.$$

Proof This follows at once from 1.6.3 and 1.5.5. □

Short alternative proofs of this theorem will appear in sections 2.4 and 6.1; however, they do not yield such a good value for the constant.

We finish this section by showing how the last result, combined with Abel summation, can be used to give an upper estimate for $\sum_{p\in P[x]}(1/p)$. (More accurate estimates, both upper and lower, will be obtained in sections 2.1 and 2.6.)

Proposition 1.6.5 *Let C_1 be as in 1.6.4. Then, for another constant c,*

$$\sum_{p\in P[x]} \frac{1}{p} \leq C_1 \log\log x + c$$

for all $x \geq 2$.

Proof Denote the sum by $S(x)$. With u_P defined as before, we have by Abel's summation formula,

$$\begin{aligned} S(x) = \sum_{n\leq x} \frac{u_P(n)}{n} &= \frac{\pi(x)}{x} + \int_2^x \frac{\pi(t)}{t^2}\, dt \\ &\leq 1 + \int_2^x \frac{C_1}{t \log t}\, dt \qquad (\text{since } \pi(x) < x) \\ &= C_1 \log\log x + c, \end{aligned}$$

where $c = 1 - C_1 \log\log 2$ (note that $\log\log 2 < 0$). □

Exercises

1 Modify the proof of 1.6.5 to show that the series $\sum_{p\in P} 1/(p \log p)$ is convergent.

2 Let C be such that $\theta(x) \le Cx$ for all $x \ge 2$. Show that

$$\sum_{p \in P[x]} \frac{\log p}{p} \le C(\log x + 1) \qquad \text{for } x \ge 2.$$

3 Write $(2n)!$ as $u_n v_n$, where

$$u_n = 1.3. \ldots .(2n-1), \qquad v_n = 2.4. \ldots .(2n).$$

Show that $u_n \le \frac{1}{2} v_n$ and hence that

$$\binom{2n}{n} \le 2^{2n-1} \qquad \text{and} \qquad \binom{2n+1}{n} \le 2^{2n}.$$

4 Show that there is a constant C such that, for all $x > 2$,

$$\sum_{p \in P,\, p > x} \frac{1}{p^2} \le \frac{C}{x \log x}.$$

1.7 Dirichlet series and the zeta function

The zeta function of a real variable

A *Dirichlet series* is a series of the form

$$\sum_{n=1}^{\infty} \frac{a(n)}{n^s}.$$

(By contrast, a *power* series is, of course, one of the form $\sum_{n=0}^{\infty} a(n) s^n$.) Any arithmetic function $a(n)$ has a corresponding Dirichlet series defined in this way. The s is to be regarded as a complex variable, and we will return shortly to the meaning and properties of n^s when s is complex.

As mentioned previously, the (Riemann) *zeta function* is the case $a(n) = 1$:

$$\zeta(s) = \sum_{n=1}^{\infty} \frac{1}{n^s}.$$

We summarize first the elementary properties of the zeta function as a function of a real variable (which we denote by σ).

Proposition 1.7.1 *The series defining $\zeta(\sigma)$ converges for all $\sigma > 1$. For such σ, $\zeta(\sigma) > 1$ and is decreasing. Also:*

$$\frac{1}{\sigma - 1} \le \zeta(\sigma) \le \frac{1}{\sigma - 1} + 1.$$

Hence:

(i) $\zeta(\sigma) \to 1$ *as* $\sigma \to \infty$,

(ii) $\zeta(\sigma) \to \infty$ *as* $\sigma \to 1^+$,

(iii) $(\sigma - 1)\zeta(\sigma) \to 1$ *as* $\sigma \to 1^+$.

Further, if $\zeta_n(\sigma) = \sum_{r=1}^{n}(1/r^\sigma)$, *then*

$$\frac{1}{(\sigma - 1)(n + 1)^{\sigma-1}} \le \zeta(\sigma) - \zeta_n(\sigma) \le \frac{1}{(\sigma - 1)n^{\sigma-1}}.$$

Proof Firstly, $\zeta(\sigma)$ is decreasing since this is true of each term $1/n^\sigma$. Now apply 1.4.4 with $f(x) = 1/x^\sigma$. The stated inequalities and statements (i), (ii) and (iii) all follow, since

$$\int_1^\infty \frac{1}{x^\sigma}\,dx = \frac{1}{\sigma - 1}, \qquad \int_n^\infty \frac{1}{x^\sigma}\,dx = \frac{1}{(\sigma - 1)n^{\sigma-1}}. \qquad \square$$

Note The function $f(x) = 1/x^\sigma$ is *convex*, hence $\frac{1}{2}f(n-1) + \frac{1}{2}f(n) \ge \int_{n-1}^{n} f(x)\,dx$ (that is, the trapezium estimate is greater than the integral; cf. 1.4.12). By adding these inequalities, we can strengthen the left-hand inequality in 1.7.1 to $\zeta(\sigma) \ge 1/(\sigma - 1) + \frac{1}{2}$.

The value of $\zeta(2)$ is $\pi^2/6$ (= 1.6449...). This can be derived from the Fourier series for x^2. Explicit values of this type for $\zeta(\sigma)$ are only known for even integers.

The relevance of the zeta function to prime numbers will be revealed in chapter 2.

The function x^s *for complex* s

We assume that the reader is familiar with the complex exponential function, which is defined by the usual series and satisfies $e^s e^w = e^{s+w}$ and $\frac{d}{ds}e^s = e^s$. For *positive, real* x and *complex* s, we define x^s (without ambiguity) to be $e^{s \log x}$, where $\log x$ means the usual real-valued logarithm of x. Clearly, $(xy)^s = x^s y^s$ and $x^s x^w = x^{s+w}$. We follow the rather strange traditional notation of analytic number theory, in which a complex variable is written as $s = \sigma + it$. With this notation, we have

$$x^s = e^{\sigma \log x} e^{it \log x} = x^\sigma[\cos(t \log x) + i \sin(t \log x)],$$

$$|x^s| = x^\sigma,$$

$$x^{\overline{s}} = \overline{x^s} \quad \text{(where } \overline{z} \text{ denotes the complex conjugate of } z\text{)}.$$

Next, we consider derivatives and integrals of x^s, first as a function of s. Since $\frac{d}{ds}e^{as} = ae^{as}$, we have

$$\frac{d}{ds}x^s = x^s \log x.$$

Now regard x^s as a function of x. This is a complex function of a real variable; for the reader who needs it, the basic principles of the calculus of such functions are set out in appendix A. In particular, the appropriate form of the chain rule applies to give (exactly as when s is real)

$$\frac{d}{dx}x^s = \frac{d}{ds}e^{s \log x} = \frac{s}{x}e^{s \log x} = sx^{s-1}.$$

The "fundamental theorem of calculus" now applies to give (again as in the real case)

$$\int_a^b x^s \, dx = \frac{1}{s+1}(b^{s+1} - a^{s+1})$$

and, for Re $s > 1$,

$$\int_a^\infty \frac{1}{x^s} \, dx = \frac{1}{(s-1)a^{s-1}}.$$

Basic properties of Dirichlet series

First, we consider the region of *absolute* convergence. As the next result shows, it is either the whole plane, empty or a half-plane.

Proposition 1.7.2 *Let $a(n)$ be a sequence. Suppose that $\sum_{n=1}^\infty (|a(n)|/n^\alpha)$ is convergent (with sum M) for some real number α. Then $\sum_{n=1}^\infty |a(n)/n^s|$ is convergent (with sum not greater than M) for all $s = \sigma + it$ with $\sigma \geq \alpha$.*

Consequently, there exists σ_a (possibly ∞ or $-\infty$) such that $\sum_{n=1}^\infty a(n)/n^s$ is absolutely convergent when $\sigma > \sigma_a$ and not absolutely convergent when $\sigma < \sigma_a$.

Proof The first statement follows from the comparison test for series, together with the inequality $|\sum_{n=1}^\infty c_n| \leq \sum_{n=1}^\infty |c_n|$, since for $\sigma \geq \alpha$,

$$\left|\frac{a(n)}{n^s}\right| = \frac{|a(n)|}{n^\sigma} \leq \frac{|a(n)|}{n^\alpha}.$$

Now let E be the set of real numbers α such that $\sum_{n=1}^\infty (|a(n)|/n^\alpha)$ is convergent, and let σ_a be the infimum of E. Let $s = \sigma + it$. If $\sigma > \sigma_a$, then (by the meaning of infimum) there exists α in E such that $\alpha < \sigma$. By the first statement, it follows that σ is in E (as required). If $\sigma < \sigma_a$, then σ is not in E. Since $|a(n)/n^s| = |a(n)|/n^\sigma$, this means that $\sum_{n=1}^\infty |a(n)/n^s|$ is not absolutely convergent. □

Proposition 1.7.3 *If $|a(n)| \leq 1$ for all n, then $\sum_{n=1}^\infty a(n)/n^s$ is absolutely convergent when* $\operatorname{Re} s > 1$ *(hence $\sigma_a \leq 1$). The same is true if $|a(n)| \leq \log n$ for all n.*

Proof The first statement follows from convergence of $\sum_{n=1}^\infty (1/n^\sigma)$. The second statement amounts to saying that $\sum_{n=1}^\infty (\log n/n^\sigma)$ is convergent. To show this, let $\sigma = 1 + 2\delta$. For large enough n, we have $\log n \leq n^\delta$, and hence $\log n/n^\sigma \leq 1/n^{1+\delta}$. This implies the fact stated. □

For the zeta function, we have $\sigma_a = 1$, since the series diverges when $\sigma = 1$. Also, we can give the following description of the location of its values.

Proposition 1.7.4 *We have $|\zeta(\sigma + it)| \leq \zeta(\sigma)$ when $\sigma > 1$. More precisely, $|\zeta(\sigma + it) - 1| \leq \zeta(\sigma) - 1$.*

Proof The first statement is immediate, and the second one is derived from $\zeta(s) - 1 = \sum_{n=2}^\infty 1/n^s$. □

This means that the values assumed by $\zeta(s)$ on a vertical line are within the circle shown.

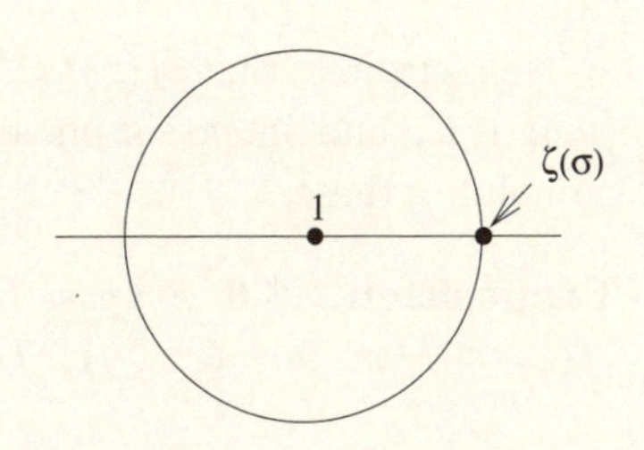

Since $\zeta(\sigma) < 2$ for $\sigma \geq 2$, it also follows that $\zeta(s) \neq 0$ (in fact, $\operatorname{Re} \zeta(s) > 0$) for $\sigma \geq 2$. (We shall improve on this statement in section 2.1).

Note also that, since $n^{\bar{s}} = \overline{n^s}$, we have $\zeta(\bar{s}) = \overline{\zeta(s)}$.

We now consider ordinary convergence (not absolute). The key is Abel summation. Restating 1.3.6 and 1.3.8 for the case $f(x) = 1/x^s$, we have:

Proposition 1.7.5 *Let $a(n)$ be a sequence, and let $A(x) = \sum_{n \leq x} a(n)$. Then for any $X \geq 1$,*

$$\sum_{n \leq X} \frac{a(n)}{n^s} = \frac{A(X)}{X^s} + s \int_1^X \frac{A(x)}{x^{s+1}} \, dx.$$

Now suppose that $s \neq 0$ and $A(x)/x^s \to 0$ as $x \to \infty$. Then if one of

$$\sum_{n=1}^{\infty} \frac{a(n)}{n^s} \qquad \text{and} \qquad s \int_1^{\infty} \frac{A(x)}{x^{s+1}}\, dx$$

converges, then so does the other, to the same value. For $X \geq 1$, we then have

$$\sum_{n>X} \frac{a(n)}{n^s} = -\frac{A(X)}{X^s} + s \int_X^{\infty} \frac{A(x)}{x^{s+1}}\, dx. \qquad \square$$

We shall call an expression of the form

$$I(s) = \int_1^{\infty} \frac{f(x)}{x^{s+1}}\, dx$$

a *Dirichlet integral.* The last result shows that every Dirichlet series is essentially a case of a Dirichlet integral. We now formulate a result on convergence for Dirichlet integrals and derive one for Dirichlet series. Of course, some integrability condition is needed when discussing Dirichlet integrals. We shall make no attempt at maximum generality here. We shall simply assume the following condition, which is obviously satisfied by summation functions $A(x)$:

(int) *f is continuous except at integers, and has left and right limits at each integer.*

This ensures that $f(x)/x^{s+1}$ is Riemann integrable on each interval $[n, n+1]$, and integrals on longer intervals are then obtained simply by combining these.

Proposition 1.7.6 *Suppose that f satisfies (int) and, for some α, we have $|f(x)| \leq Mx^{\alpha}$ for all $x \geq 1$. Then*

$$\int_1^{\infty} \frac{f(x)}{x^{s+1}}\, dx$$

is convergent for all $s = \sigma + it$ with $\sigma > \alpha$. Denote its value by $I(s)$, and let

$$\int_1^{X} \frac{f(x)}{x^{s+1}}\, dx = I_X(s).$$

Then, for $\sigma > \alpha$,

$$|I(s)| \leq \frac{M}{\sigma - \alpha} \qquad \text{and} \qquad |I(s) - I_X(s)| \leq \frac{M}{(\sigma - \alpha)X^{\sigma - \alpha}}.$$

Proof We have

$$\left|\frac{f(x)}{x^{s+1}}\right| \le \frac{M}{x^{\sigma-\alpha+1}}.$$

Now, for $X \ge 1$,

$$\int_X^\infty \frac{1}{x^{\sigma-\alpha+1}}\,dx = \frac{1}{(\sigma-\alpha)X^{\sigma-\alpha}}.$$

The statements follow, by the inequality $|\int_a^b f(x)\,dx| \le \int_a^b |f(x)|\,dx$ (for a proof of this inequality when $f(x)$ is complex, see appendix A). □

Proposition 1.7.7 *Let $a(n)$ be a sequence, and let $A(x) = \sum_{n\le x} a(n)$. Suppose that $|A(x)| \le Mx^\alpha$ for all $x \ge 1$, where $\alpha \ge 0$. Then $\sum_{n=1}^\infty a(n)/n^s$ is convergent for all $s = \sigma + it$ with $\sigma > \alpha$. Denote its sum by $F(s)$, and let $\sum_{n\le X} a(n)/n^s = F_X(s)$. Then*

$$|F(s)| \le M\frac{|s|}{\sigma-\alpha} \qquad \textit{and} \qquad |F(s) - F_X(s)| \le \frac{M}{X^{\sigma-\alpha}}\left(\frac{|s|}{\sigma-\alpha}+1\right).$$

Proof This follows at once by combining 1.7.5 and 1.7.6. Note that $|A(x)/x^s| \le M/x^{\sigma-\alpha}$, which tends to 0 as $x \to \infty$. □

Remark In particular, if $|A(x)| \le M$ for all x, then $|F(s)| \le M|s|/\sigma$. The ratio $|s|/\sigma$ has a simple geometrical meaning: it equals $\sec\theta$ when s is expressed as $re^{i\theta}$.

We can now contrast absolute and non-absolute convergence. Recall that a *power* series converges absolutely within the circle where it converges at all. The situation is different for Dirichlet series.

Proposition 1.7.8 *Suppose that $\sum_{n=1}^\infty a(n)/n^\alpha$ is convergent for a certain real α. Then $\sum_{n=1}^\infty a(n)/n^s$ is convergent for all $s = \sigma + it$ with $\sigma > \alpha$.*

Consequently, there exists σ_c (possibly ∞ or $-\infty$) such that the series is convergent for all $s = \sigma + it$ with $\sigma > \sigma_c$ and not convergent for $\sigma < \sigma_c$. Further, $\sigma_a \le \sigma_c + 1$.

Proof Let $b(n) = a(n)/n^\alpha$ and $B(x) = \sum_{n\le x} b(n)$. Then $\sum_{n=1}^\infty b(n)$ is convergent, so there exists M such that $|B(x)| \le M$ for all $x \ge 1$. By the case $\alpha = 0$ in 1.7.7, $\sum_{n=1}^\infty b(n)/n^s$ is convergent when Re $s > 0$. But

$$\frac{a(n)}{n^s} = \frac{b(n)}{n^{s-\alpha}},$$

so $\sum_{n=1}^\infty a(n)/n^s$ is convergent when $\sigma > \alpha$.

The existence of σ_c now follows in the same way as for σ_a. With α as above, we show that $\sum_{n=1}^{\infty} a(n)/n^s$ is absolutely convergent when $\sigma > \alpha + 1$: it then follows that $\sigma_a \leq \sigma_c + 1$. Now $|b(n)|$ is bounded, say by K, so

$$\left|\frac{a(n)}{n^s}\right| = \frac{|b(n)|}{|n^{s-\alpha}|} \leq \frac{K}{n^{\sigma-\alpha}}.$$

Since $\sigma - \alpha > 1$, the series $\sum_{n=1}^{\infty} |a(n)/n^s|$ is convergent, as required. □

A further easy application of Abel summation shows that when σ_c is positive, it is actually characterized by the condition in 1.7.7. We use the O notation (see appendix E).

Proposition 1.7.9 *Suppose that $\sum_{n=1}^{\infty} a(n)/n^\sigma$ converges for a certain $\sigma > 0$. Then $A(x) = O(x^\sigma)$ for $x \geq 1$.*

Let α be the infimum of the numbers σ such that $A(x) = O(x^\sigma)$ for $x \geq 1$. If $\alpha > 0$, then $\sigma_c = \alpha$.

Proof Let $b(n) = a(n)/n^\sigma$, so that $\sum_{n=1}^{\infty} b(n)$ is convergent. In particular, $|B(x)|$ is bounded, say by M. By Abel's summation formula,

$$A(x) = \sum_{n \leq x} n^\sigma b(n) = x^\sigma B(x) - \sigma \int_1^x t^{\sigma-1} B(t)\, dt.$$

Hence

$$|A(x)| \leq Mx^\sigma + M \int_1^x \sigma t^{\sigma-1}\, dt \leq 2Mx^\sigma.$$

With α as stated, let $\sigma > \alpha$. Take σ_1 such that $\alpha < \sigma_1 < \sigma$. Then $A(x) = O(x^{\sigma_1})$ for $x \geq 1$, so by 1.7.7, $\sum_{n=1}^{\infty} a(n)/n^\sigma$ is convergent. So $\sigma_c \leq \alpha$.

Now let $0 < \sigma < \alpha$. Then $A(x)$ is not $O(x^\sigma)$, so by the above, $\sum_{n=1}^{\infty} a(n)/n^\sigma$ is divergent. Hence $\sigma_c \geq \alpha$. □

Clearly, if $\alpha = 0$, then $\sigma_c \leq 0$. When $\sigma_c < 0$, it can be characterized using the *tail* of the series $\sum_{n=1}^{\infty} a(n)$: see exercise 5.

If $a(n) \geq 0$ for all n, then absolute convergence of $\sum_{n=1}^{\infty} a(n)/n^\sigma$ (for real σ) coincides with convergence. Since σ_c is the infimum of *real* σ for which the series converges (and similarly for σ_a), it follows that $\sigma_a = \sigma_c$ in this case. In particular, for the zeta function, we have $\sigma_a = \sigma_c = 1$, since the series diverges when $\sigma = 1$.

Example Let $a(n) = (-1)^{n-1}$. Then $A(n)$ is bounded, since it is alternately 1 and 0. So, by 1.7.7, the series converges whenever $\sigma > 0$. It does not converge when $\sigma = 0$, so $\sigma_c = 0$. Since $|a(n)| = 1$ for all n, we have $\sigma_a = 1$.

It is easy to express the sum $F(s)$ of this series in terms of $\zeta(s)$ for Re $s > 1$:

$$\zeta(s) - F(s) = 2\sum_{n=1}^{\infty} \frac{1}{(2n)^s} = \frac{2}{2^s}\zeta(s),$$

so $F(s) = (1 - 2^{1-s})\zeta(s)$.

Note Let $f(x)$ be a function defined on $(0, \infty)$. The *Mellin transform* of f is the function defined by

$$M(s) = \int_0^{\infty} x^{s-1} f(x)\, dx$$

where this converges. For functions that are zero on $(0, 1)$, the Mellin transform coincides with our "Dirichlet integral", with $-s$ substituted for s. But a general study of Mellin transforms requires consideration of convergence of the integral on $(0, 1)$ as well as $[1, \infty)$.

Differentiability

It remains to establish a very important property of functions defined by Dirichlet series, namely that they are *holomorphic*, that is, differentiable in the sense of complex functions. We have seen that the derivative of $a(n)n^{-s}$ is $-a(n)n^{-s} \log n$. The point is to show that termwise differentiation of the series is valid. The key notion here is *uniform convergence*. Recall that if f_n $(n \geq 1)$ and f are (complex) functions and E is a set in the complex plane, then (f_n) is said to converge to f *uniformly on* E if, for any $\varepsilon > 0$, there exists n_0 such that, for all $n \geq n_0$, one has $|f_n(s) - f(s)| \leq \varepsilon$ for all $s \in E$. If E is the closed rectangle consisting of $s = \sigma + it$ with $\sigma_1 \leq \sigma \leq \sigma_2$, $t_1 \leq t \leq t_2$, its *interior*, int(E), is the set given by $\sigma_1 < \sigma < \sigma_2$, $t_1 < t < t_2$. We assume the following standard theorem from complex analysis:

> *If (f_n) is a sequence of holomorphic functions that converges to a function f uniformly on some rectangle E, then f is holomorphic on* int(E) *and* $f'(s) = \lim_{n\to\infty} f_n'(s)$ *for all s in* int(E).

(We remind the reader that a similar statement for *real* functions is not true!)

Proposition 1.7.10 *Suppose that $\sum_{n=1}^{\infty} a(n)/n^s$ converges to $F(s)$ for* Re $s > \sigma_c$. *Then $F(s)$ is holomorphic for such s, with derivative given by*

$$F'(s) = -\sum_{n=1}^{\infty} \frac{a(n) \log n}{n^s}.$$

Proof Choose $\alpha > \sigma_c$. As in 1.7.8, let $b(n) = a(n)/n^\alpha$. Then $|B(N)| \le M$ (say) for all $N \ge 1$, and $F(s) = G(s - \alpha)$, where $G(s) = \sum_{n=1}^\infty b(n)/n^s$. We will show that $G(s)$ is holomorphic, with derivative as stated, for Re $s > 0$. It then follows that

$$F'(s) = G'(s-\alpha) = -\sum_{n=1}^\infty \frac{b(n)\log n}{n^{s-\alpha}} = \sum_{n=1}^\infty \frac{a(n)\log n}{n^s}$$

whenever Re $s > \alpha$, and hence in fact whenever Re $s > \sigma_c$.

Write $G_N(s) = \sum_{n=1}^N b(n)/n^s$. Since this is a finite sum, we have

$$G_N'(s) = -\sum_{n=1}^N \frac{b(n)\log n}{n^s}.$$

Take $\delta > 0$, $R > 0$ and let $E_{\delta,R}$ be the semi-infinite rectangle consisting of $s = \sigma + it$ with $\sigma \ge \delta$ and $|t| \le R$. We will show that $G_N(s) \to G(s)$ uniformly on $E_{\delta,R}$. The statement then follows for all s in $\mathrm{int}(E_{\delta,R})$. But every s with $\sigma > 0$ belongs to such a set for a suitable choice of δ and R, so the result holds for all such s.

By 1.7.7 (the case $\alpha = 0$), we have

$$|G(s) - G_N(s)| \le \frac{M}{N^\sigma}\left(\frac{|s|}{\sigma} + 1\right).$$

Now, for all s in $E_{\delta,R}$,

$$\frac{|s|}{\sigma} \le \frac{\sigma + |t|}{\sigma} = 1 + \frac{|t|}{\sigma} \le 1 + \frac{R}{\delta},$$

and hence

$$|G(s) - G_N(s)| \le \frac{M}{N^\delta}\left(2 + \frac{R}{\delta}\right),$$

which tends to 0 as $N \to \infty$. This proves the required uniform convergence. □

Remark Uniform convergence is rather easier if we are interested only in $\sigma > \sigma_a$. For if $\alpha > \sigma_a$, then the inequality appearing in the proof of 1.7.2, together with the "M-test", shows that the series converges uniformly on the set of s such that $\sigma \ge \alpha$.

We will also need the fact that functions defined by Dirichlet *integrals*, under the conditions of 1.7.6, are holomorphic (though of course this is now clear when the Dirichlet integral is obtained from a Dirichlet series as in 1.7.5). The uniform convergence step is easy, by the estimate in 1.7.6.

However, we now need information about the derivative (with respect to s) of the integral

$$I_X(s) = \int_1^X \frac{f(x)}{x^{s+1}}\, dx$$

instead of a finite sum $\sum_{n=1}^N a(n)/n^s$. In place of differentiation of a finite sum, we need to know that "differentiation under the integral sign" is valid, giving

$$I_X'(s) = -\int_1^X \frac{f(x)\log x}{x^{s+1}}\, dx.$$

There is indeed a general theorem ensuring that this is valid. It is not very hard to give a proof specialized to the case we want: this is done in appendix D. Assuming this result, we deduce:

Proposition 1.7.11 *Suppose that $f(x)$ satisfies (int) and $|f(x)| \le Mx^\alpha$ for all $x \ge 1$. Let*

$$I(s) = \int_1^\infty \frac{f(x)}{x^{s+1}}\, dx.$$

Then $I(s)$ is holomorphic for $s = \sigma + it$ with $\sigma > \alpha$, and

$$I'(s) = -\int_1^\infty \frac{f(x)\log x}{x^{s+1}}\, dx.$$

Proof We only need to check uniform convergence. Let $E_\delta = \{s : \sigma \ge \alpha+\delta\}$. By 1.7.6, for $s \in E_\delta$ and integers N,

$$|I(s) - I_N(s)| \le \frac{M}{(\sigma-\alpha)N^{\sigma-\alpha}} \le \frac{M}{\delta N^\delta}.$$

Hence $I_N(s) \to I(s)$ as $N \to \infty$, uniformly on E_δ. □

In particular, for the zeta function, we have

$$\zeta'(s) = -\sum_{n=1}^\infty \frac{\log n}{n^s} = -\sum_{n=2}^\infty \frac{\log n}{n^s}$$

when Re $s > 1$. We finish this section with a rough estimation of this function. Clearly, $|\zeta'(s)| \le -\zeta'(\sigma)$, where $s = \sigma + it$.

Proposition 1.7.12 *For all $\sigma > 1$, we have*

$$-\zeta'(\sigma) \le \frac{1}{(\sigma-1)^2} + \frac{3}{4}.$$

Proof We assume the integral

$$\int_1^\infty \frac{\log x}{x^\sigma}\,dx = \frac{1}{(\sigma-1)^2},$$

which is readily established by integrating by parts on $[1,X]$ and considering the limit as $X \to \infty$. We use the integral to estimate the series, but this is not quite as simple as before because $\log x/x^\sigma$ does not decrease throughout $[1,\infty)$. In fact, differentiation shows that it decreases when $\log x \geq 1/\sigma$. Hence (for any $\sigma > 1$) it certainly decreases for all $x \geq 3$, so that

$$\sum_{n=4}^\infty \frac{\log n}{n^\sigma} \leq \int_3^\infty \frac{\log x}{x^\sigma}\,dx < \frac{1}{(\sigma-1)^2}.$$

The terms for $n = 2, 3$ contribute less than $\frac{1}{2}\log 2 + \frac{1}{3}\log 3 \approx 0.7128 < \frac{3}{4}$. (If $\sigma \geq \frac{3}{2}$, then $\log x/x^\sigma$ decreases for all $x \geq 2$, so $\frac{3}{4}$ can be replaced by the term $\log 2/2^\sigma$. Actually, one can show that the $\frac{3}{4}$ is not needed at all.) □

Exercises

1 By reversing the order of summation, prove that $\sum_{m=2}^\infty [\zeta(m)-1] = 1$, and obtain a series expression for $\sum_{m=2}^\infty \zeta'(m)$.

2 By using integral estimation for $\sum_{n=3}^\infty 1/n^\sigma$, show that

$$\zeta(\sigma) \leq 1 + \frac{\sigma+1}{\sigma-1}\frac{1}{2^\sigma}.$$

3 Let $A(x) = \sum_{n\leq x} a(n)$, and suppose that $|A(x)| \leq Mx^\alpha$ for all $x \geq 1$, where $\alpha > 0$. Write $b(n) = a(n)\log n$. Use Abel summation to show that $|B(x)| \leq Mx^\alpha(\log x + \alpha^{-1})$, and deduce that $\sum_{n=1}^\infty b(n)/n^s$ is convergent when $\sigma > \alpha$. Adapt this for the case $\alpha = 0$.

4 (*An estimate in terms of t*) Suppose that $|a(n)| \leq m$ and $|A(n)| \leq M$ for all n. Let $F(s) = \sum_{n=1}^\infty a(n)/n^s$ where it converges. Let $t \geq 0$ be given. By taking $X = t + 2$ in 1.7.7, show that

$$|F(1+it)| \leq m\log(t+2) + M + m.$$

5 (*Companion to 1.7.7 in terms of the tail of* $\sum a(n)$) Suppose that $\sum_{n=1}^\infty a(n)$ is convergent (say to L), and write $R(x) = \sum_{n>x} a(n)$. Suppose that for some M and some $\alpha > 0$, we have $|R(x)| \leq Mx^{-\alpha}$ for all $x \geq 1$. By substituting $L - R(x)$ for $A(x)$ in Abel's summation formula, show that if Re $s < \alpha$, then $\sum_{n=1}^\infty a(n)n^s$ converges to $L + s\int_1^\infty x^{s-1}R(x)\,dx$. Deduce that $\sigma_c \leq -\alpha$.

6 Suppose that $a(n) \geq 0$ for all n. Suppose also that $\sum_{n=1}^{\infty} a(n)/n^s$ converges to $F(s)$ for all real $s > 0$ and that $F(s) \to L$ as $s \to 0^+$. Prove that $\sum_{n=1}^{\infty} a(n)$ converges to L. Deduce the following variant: if $F(s) \to L_1$ as $s \to 1^+$, then $\sum_{n=1}^{\infty} a(n)/n = L_1$.

7 Suppose that $B(x)$ is bounded and tends to L as $x \to \infty$. Let

$$G(s) = s \int_1^{\infty} \frac{B(x)}{x^{s+1}} \, dx.$$

In this integral, replace $B(x)$ by $B(x) - L$, and split $[1, \infty)$ into $[1, X]$ and $[X, \infty)$ to show that $G(s) \to L$ as $s \to 0^+$ (with s real).

Now let $F(s) = \sum_{n=1}^{\infty} a(n)/n^s$, and suppose that $A(x) - \log x \to L$ as $x \to \infty$. Using the identity $s \int_1^{\infty} x^{-s-1} \log x \, dx = 1/s$, deduce that

$$F(s) - \frac{1}{s} \to L \quad \text{as } s \to 0^+.$$

Deduce in particular that $\zeta(s) - 1/(s-1) \to \gamma$ as $s \to 1^+$.

8 The *gamma function* is defined, for Re $s > 0$, by

$$\Gamma(s) = \int_0^{\infty} x^{s-1} e^{-x} \, dx.$$

By substituting $x = ny$ in this expression and assuming validity of termwise integration of the relevant series, show that, for Re $s > 1$,

$$\Gamma(s)\zeta(s) = \int_0^{\infty} \frac{x^{s-1}}{e^x - 1} \, dx$$

9 (*Uniqueness and zeros*) Suppose that $\sum_{n=1}^{\infty} a(n)/n^s$ converges to $F(s)$ when Re $s > \sigma_c$, and that there is a sequence $s_k = \sigma_k + it_k$, with $\sigma_k \to \infty$ as $k \to \infty$, such that $F(s_k) = 0$ for all k. Show by the following steps that $a(n) = 0$ for all n. Supposing the opposite, let N be the first n such that $a(n) \neq 0$. Fix $c > \sigma_a$. Show that there is a constant B such that, for $\sigma > c$,

$$\sum_{n=N+1}^{\infty} \frac{|a(n)|}{n^{\sigma}} \leq \frac{B}{(N+1)^{\sigma - c}},$$

and hence deduce that $a(N) = 0$. Deduce further: (i) if $a(n) \neq 0$ for some n, then there exists σ_1 such that $F(s) \neq 0$ whenever $\sigma > \sigma_1$; (ii) if $\sum_{n=1}^{\infty} a(n)/n^{\sigma} = \sum_{n=1}^{\infty} b(n)/n^{\sigma}$ for all large enough σ, then $a(n) = b(n)$ for all n.

1.8 Convolutions

Given an arithmetic function $a(n)$, let $F_a(s)$ be the function defined by the corresponding Dirichlet series (where it converges):

$$F_a(s) = \sum_{n=1}^{\infty} \frac{a(n)}{n^s}.$$

Sometimes $F_a(s)$ is called the *generating function* for $a(n)$. The underlying idea is that we might be able to deduce information about $a(n)$ from the behaviour of $F_a(s)$. This is in fact the process that will eventually lead to a proof of the prime number theorem.

We list some examples of Dirichlet series corresponding to arithmetic functions, using our previous notation from section 1.2:

$$F_u(s) = \zeta(s), \quad F_{e_1}(s) = 1, \quad F_{e_n}(s) = \frac{1}{n^s}, \quad F_{u_P}(s) = \sum_{p \in P} \frac{1}{p^s},$$

and if $a = u_S$, where S is the set of squares, then

$$F_a(s) = \sum_{n=1}^{\infty} \frac{1}{(n^2)^s} = \zeta(2s) \quad \text{for } \operatorname{Re} s > \tfrac{1}{2}.$$

Now consider the product of two Dirichlet series. We use the following result about multiplication of series (the reader may be content to assume it, but a proof is given in appendix B).

(MS) *If $\sum_{m=1}^{\infty} a_m$ and $\sum_{n=1}^{\infty} b_n$ are absolutely convergent (with sums A and B respectively), then however the terms $a_m b_n$ are arranged as a single series, the resulting series converges to AB.*

Consider the product

$$F_a(s) F_b(s) = \left(\sum_{j=1}^{\infty} \frac{a(j)}{j^s} \right) \left(\sum_{k=1}^{\infty} \frac{b(k)}{k^s} \right).$$

In the range where both series are absolutely convergent, we apply (MS) and collect together, for each n, the terms

$$\frac{a(j)b(k)}{j^s k^s}$$

with $jk = n$. The conclusion is that the product equals $\sum_{n=1}^{\infty} c(n)/n^s$, where

$$c(n) = \sum_{jk=n} a(j)b(k) = \sum_{j|n} a(j) b\left(\frac{n}{j}\right).$$

Here $\sum_{jk=n}$ means the sum taken over all ordered pairs (j,k) with $jk = n$ and $\sum_{j|n}$ means the sum over all (positive) divisors j of n. The function c defined in this way is called the *Dirichlet convolution* (or just *convolution*) of a and b, and is denoted by $a*b$. (It is also sometimes called the "Dirichlet product").

The above reasoning shows that multiplication of Dirichlet series translates into convolution of the sequences of coefficients. The precise statement is

Proposition 1.8.1 *Suppose that* $F_a(s) = \sum_{n=1}^{\infty} a(n)/n^s$ *and* $F_b(s) = \sum_{n=1}^{\infty} b(n)/n^s$*, both series being absolutely convergent for* Re $s > \sigma_0$*. Then, for such* s*,*

$$F_a(s)F_b(s) = F_{a*b}(s). \qquad \square$$

Actually, this statement is correct provided that *one* of the two series is absolutely convergent (see appendix B again).

One could, of course, define convolutions without mentioning Dirichlet series, but the definition is motivated by the partnership between arithmetic functions and Dirichlet series, and 1.8.1 will point the way to a number of useful convolution identities.

Note When products of *power* series $\sum_{n=0}^{\infty} a(n)z^n$ are considered, the resulting coefficients are $\sum_{j=0}^{n} a(j)b(n-j)$. This defines another type of convolution of a and b. However, in the context of the present subject, the word "convolution" always means Dirichlet convolution.

Obviously, $(a*b)(1) = a(1)b(1)$. Note also that

$$(a*u)(n) = \sum_{j|n} a(j).$$

The next result lists the elementary algebraic properties of convolutions. They correspond to familiar properties of multiplication.

Proposition 1.8.2 *For arithmetic functions a,b,c, we have*

(i) $a*b = b*a$,

(ii) $a*(b+c) = a*b + a*c$,

(iii) $a*e_1 = a$ *(so* e_1 *acts as an identity),*

(iv) $a*(b*c) = (a*b)*c$.

Proof (i), (ii) Obvious.

(iii) Since $e_1(j) = 0$ for $j > 1$, we have $(a*e_1)(n) = a(n)e_1(1) = a(n)$.

(iv) We have

$$\begin{aligned}[a*(b*c)](n) &= \sum_{im=n} a(i)(b*c)(m)\\ &= \sum_{im=n} a(i) \sum_{jk=m} b(j)c(k)\\ &= \sum_{ijk=n} a(i)b(j)c(k).\end{aligned}$$

Similarly, $[(a*b)*c](n)$ equates to the same expression. □

We mention some examples and special cases.

Proposition 1.8.3 *We have* $u*u=\tau$ *and* $u_P*u=\omega$.

Proof First, $(u*u)(n)=\sum_{j|n}u(j)=\sum_{j|n}1=\tau(n)$. Second, $(u_P*u)(n)=\sum_{j|n}u_P(j)$. This is the number of prime divisors of n, that is, $\omega(n)$. □

Note the corresponding series identities (both valid for Re $s>1$):

$$\sum_{n=1}^{\infty}\frac{\tau(n)}{n^s}=\zeta(s)^2, \qquad \sum_{n=1}^{\infty}\frac{\omega(n)}{n^s}=\zeta(s)\sum_{p\in P}\frac{1}{p^s}.$$

(A direct proof that $\sum_{n=1}^{\infty}\tau(n)/n^s$ converges when Re $s>1$ would require 1.4.5 and 1.7.7.)

We list some further examples, all easily verified; the corresponding series identities are obvious:

$$e_j*e_k=e_{jk},$$

$$(e_j*u)(n)=\begin{cases}1 & \text{if } j|n,\\ 0 & \text{otherwise,}\end{cases}$$

$$[(e_1-2e_2)*u](n)=(-1)^{n-1}.$$

More examples will follow later. Indeed, many of the relationships of number theory can be expressed neatly and concisely as convolution identities.

Next, we consider sums of terms $(a*b)(n)$. Clearly, we have

$$\sum_{n=1}^{\infty}(a*b)(n)=\left(\sum_{n=1}^{\infty}a(n)\right)\left(\sum_{n=1}^{\infty}b(n)\right)$$

if both series on the right are absolutely convergent: this is simply the case $s=0$ in 1.8.1. As usual, we are more interested in the partial sum $\sum_{n\le x}(a*b)(n)$. The next result gives some expressions for it.

Proposition 1.8.4 *Let a, b be arithmetic functions. Let* $A(x) = \sum_{n\leq x} a(n)$, $B(x) = \sum_{n\leq x} b(n)$. *Then*

$$\sum_{n\leq x}(a*b)(n) = \sum_{jk\leq x} a(j)b(k) = \sum_{j\leq x} a(j)B\left(\frac{x}{j}\right) = \sum_{k\leq x} A\left(\frac{x}{k}\right) b(k).$$

Proof By the definition, we have

$$\sum_{n\leq x}(a*b)(n) = \sum_{n\leq x}\sum_{jk=n} a(j)b(k) = \sum_{jk\leq x} a(j)b(k).$$

We can rewrite this sum as

$$\sum_{j\leq x} a(j) \sum_{k\leq x/j} b(k) = \sum_{j\leq x} a(j)B\left(\frac{x}{j}\right).$$

The second expression is obtained by interchanging a and b. □

Corollary 1.8.5 *For any arithmetic function a,*

$$\sum_{n\leq x}(a*u)(n) = \sum_{j\leq x} a(j)\left[\frac{x}{j}\right] = \sum_{k\leq x} A\left(\frac{x}{k}\right).$$

Proof Take $b = u$, so that $B(x) = [x]$. □

In the light of 1.8.3, the identities in 1.2.2 are a special case of the first equality. In fact, the method of 1.2.2 gives a direct route to this equality, as follows. For fixed j, there is a term $a(j)$ for each multiple rj not greater than x; clearly, there are $[x/j]$ such multiples.

A further rewriting of the sum in 1.8.4 expresses it in terms of shorter sums, which is sometimes very effective. A typical choice for the y appearing in the statement is $x^{1/2}$.

Proposition 1.8.6 (Dirichlet's hyperbola identity) *If* $1 < y < x$, *then*

$$\sum_{n\leq x}(a*b)(n) = \sum_{j\leq y} a(j)B\left(\frac{x}{j}\right) + \sum_{k\leq x/y} b(k)A\left(\frac{x}{k}\right) - A(y)B\left(\frac{x}{y}\right).$$

Proof We have seen that the sum equates to $S = \sum_{jk\leq x} a(j)b(k)$. Let S_1 and S_2, respectively, be the sum of all such terms with $j \leq y$ and with $k \leq x/y$. Then, as in 1.8.4,

$$S_1 = \sum_{j\leq y} a(j)B(x/j), \qquad S_2 = \sum_{k\leq x/y} b(k)A(x/k).$$

The sum $S_1 + S_2$ counts twice the terms with both $j \leq y$ and $k \leq x/y$. The sum of these is $A(y)B(x/y)$, so the statement follows. It helps to picture this argument geometrically: we are counting "lattice points" (j,k) lying below the hyperbola $\xi\eta = x$ in the (ξ,η) plane. □

The product rule for differentiation translates nicely into a corresponding rule for convolutions, involving the function $\log n$. Since

$$F'_a(s) = -\sum_{n=1}^{\infty} \frac{a(n)\log n}{n^s},$$

$F'_a(s)$ is the Dirichlet series corresponding to $-\ell a$, where $\ell(n) = \log n$. The product rule equates $(F_aF_b)'$ to $F'_aF_b + F_aF'_b$. It is easy to verify the convolution identity suggested by this:

Proposition 1.8.7 *Write $\ell(n) = \log n$. For any arithmetic functions a,b, we have*

$$\ell(a * b) = (\ell a) * b + a * (\ell b).$$

Proof For each n,

$$\begin{aligned}(\log n)(a*b)(n) &= \log n \sum_{jk=n} a(j)b(k)\\ &= \sum_{jk=n} (\log j + \log k)a(j)b(k)\\ &= [(\ell a) * b](n) + [a * (\ell b)](n).\end{aligned}$$

□

We finish with some facts relating to multiplicative functions.

Proposition 1.8.8 *If a and b are multiplicative, then so is $a * b$.*

Proof Let $(m,n) = 1$. We assume the following fact, which follows at once from unique prime factorization (or else can be proved by elementary number theory): every divisor r of mn is uniquely expressible as jk, where $j|m$ and $k|n$. Clearly, we then have $(j,k) = (m/j, n/k) = 1$. Hence

$$\begin{aligned}(a*b)(mn) &= \sum_{r|mn} a(r)b\left(\frac{mn}{r}\right)\\ &= \sum_{j|m}\sum_{k|n} a(jk)b\left(\frac{m}{j}\frac{n}{k}\right)\\ &= \sum_{j|m} a(j)b\left(\frac{m}{j}\right)\sum_{k|n} a(k)b\left(\frac{n}{k}\right)\\ &= (a*b)(m)\,(a*b)(n).\end{aligned}$$

□

The fact that τ $(= u * u)$ is multiplicative is a special case of this result. A limited distributive law holds for pointwise products, as follows.

Proposition 1.8.9 *If a,b,c are arithmetic functions and c is completely multiplicative, then* $(ac) * (bc) = (a * b)c$.

Proof For each n,

$$[(ac) * (bc)](n) = \sum_{jk=n} (ac)(j)(bc)(k) = \sum_{jk=n} a(j)c(j)b(k)c(k).$$

Since $c(j)c(k) = n$, this equates to

$$c(n) \sum_{jk=n} a(j)b(k) = c(n)(a * b)(n). \qquad \square$$

Example Taking $c(n) = 1/n$ in 1.8.9, we see that if $a_1(n) = a(n)/n$ and $b_1(n) = b(n)/n$, then $(a_1 * b_1)(n) = (a * b)(n)/n$.

Exercises

1. Define $\sigma_r(n)$ to be $\sum_{j|n} j^r$. Express σ_r as a convolution. Deduce that it is multiplicative, and express $\sum_{n=1}^{\infty} \sigma_r(n)/n^s$ in terms of the zeta function for suitable s. Show that $\sigma_{-1}(n) = \sigma_1(n)/n$.
2. Describe $u_P * u_P$ by giving its values for n of the form (i) pq (where p, q are prime), (ii) p^2, (iii) neither pq nor p^2.
3. Define an arithmetic function a such that $a * u = \Omega$.
4. Show that if a is any arithmetic function with $a(1) \neq 0$, then a has a unique inverse b with respect to convolution. (For each n, show how to express $b(n)$ as a combination of earlier terms.)
5. Suppose that $\sum_{n=1}^{\infty} |a(n)| \log n$ is convergent and $\sum_{n=1}^{\infty} a(n) = 0$. Prove that $\sum_{n=1}^{\infty} (a * a)(n) \log n = 0$. Relate this to the corresponding Dirichlet series.
6. Apply Dirichlet's hyperbola identity, with $y = x^{1/2}$, to show that if $A(x)$ and $B(x)$ (hence also $a(n)$ and $b(n)$) are $O(x^{\alpha})$ for $x \geq 1$, where $\alpha \geq 0$, and $c = a * b$, then $C(x)$ is $O(x^{\alpha+1/2})$ for $x \geq 1$ (so that the series $\sum_{n=1}^{\infty} c(n)/n^s$ is convergent for Re $s > \alpha + \frac{1}{2}$).

2
Some important Dirichlet series and arithmetic functions

This chapter opens with the "Euler product", a remarkable identity that unlocks a whole vista of further ideas. Stated in modern terms, it says that, for a completely multiplicative function $a(n)$,

$$\sum_{n=1}^{\infty} a(n) = \prod_{p \in P} [1 - a(p)]^{-1}.$$

In other words, an infinite series is equated to a product involving only the *primes*. This identity encapsulates in analytic form the fact that integers are uniquely expressible as products of primes. At a stroke, applied to $a(n) = n^{-s}$, it relates the zeta function to prime numbers. It is also a fertile source of further Dirichlet series and associated convolutions. In particular, it leads to the important Dirichlet series

$$\frac{1}{\zeta(s)} = \sum_{n=1}^{\infty} \frac{\mu(n)}{n^s}, \qquad \frac{\zeta'(s)}{\zeta(s)} = -\sum_{n=1}^{\infty} \frac{\Lambda(n)}{n^s},$$

in which μ is the *Möbius function* and Λ is the *von Mangoldt* function. Corresponding convolution identities are then easily checked directly. The definitions of these functions seem a little strange if presented without motivation, but the Euler product leads us to them in a natural way.

The summation function of $\Lambda(n)$ (denoted by $\psi(x)$) is similar to Chebyshev's $\theta(x)$, but differs by counting powers of primes as well as primes. We show that it would do just as well for the purpose of proving the prime number theorem. The final step, which must wait until chapter 3, will be to deduce the required estimation of $\psi(x)$ from the behaviour of the function $\zeta'(s)/\zeta(s)$.

Meanwhile, the methods of the present chapter are enough to give very satisfactory estimates of the summation functions of various other arithmetic functions and of sums like $\sum_{p \in P[x]} 1/p$. A number of results of this sort are described in sections 2.5 and 2.6. The reader wanting a fast track to the prime number theorem can defer these sections.

2.1 The Euler product

We now present a remarkable identity discovered by Euler as long ago as 1737. This identity, the "Euler product", will serve to explain the connection between the zeta function and prime numbers. It will also enable us to extend greatly our repertoire of Dirichlet series and associated convolutions. As before, we denote by P the set of all prime numbers, and by $P[N]$ the set of prime numbers not greater than N. When considering series of the form $\sum_{p \in P} f(p)$, it is understood that the primes are taken in increasing order.

We again use the following result about multiplication of series (see appendix B):

(MS) *If $\sum_{m=1}^{\infty} a_m$ and $\sum_{n=1}^{\infty} b_n$ are absolutely convergent (with sums A and B respectively), then however the terms $a_m b_n$ are arranged as a single series, the resulting series converges to AB.*

We also assume at this point that the reader is familiar with the notion of infinite products. For those who are not, the essential facts required are given in appendix C. In particular, we assume the following theorem:

(PR) *If $\sum_{n=1}^{\infty} |a_n|$ is convergent (where the a_n are real or complex), then the product $\prod_{n=1}^{\infty}(1 + a_n)$ is convergent to a non-zero value.*

Euler's idea, slightly generalized, is as follows. Suppose that $a(n)$ is completely multiplicative and $\sum_{n=1}^{\infty} |a(n)|$ is convergent. For each prime p, we have the geometric series

$$[1 - a(p)]^{-1} = 1 + a(p) + a(p)^2 + \cdots.$$

This series converges, since $a(p^k) = a(p)^k$ and by assumption $\sum_{k=1}^{\infty} a(p^k)$ is convergent. Also, $\sum_{p \in P} |a(p)|$ is convergent, so, by (PR), the product $\prod_{p \in P}[1 - a(p)]^{-1}$ is convergent. Consider the corresponding "formal" product of geometric series. One term is 1, which equals $a(1)$, since a is completely multiplicative. Each integer $n > 1$ is a unique product of primes, say $n = p_1^{k_1} \dots p_r^{k_r}$. Since a is completely multiplicative, $a(n) = a(p_1)^{k_1} \dots a(p_r)^{k_r}$. This term appears exactly once in the above product of geometric series, suggesting that this product equals $\sum_{n=1}^{\infty} a(n)$. We now prove that this is true.

Theorem 2.1.1 (the generalized Euler product) *Suppose that $a(n)$ is a completely multiplicative arithmetic function such that $\sum_{n=1}^{\infty} |a(n)|$ is convergent. Then $\sum_{n=1}^{\infty} a(n) \neq 0$, and*

$$\sum_{n=1}^{\infty} a(n) = \prod_{p \in P} \frac{1}{1 - a(p)}.$$

Proof Note first that, by (PR), the product $\prod_{p \in P}[1 - a(p)]$ converges to a non-zero value, and hence the same is true of $\prod_{p \in P}[1 - a(p)]^{-1}$.

Now fix N. Let E_N be the set of all positive integers (including 1) expressible as products of powers of the primes in $P[N]$, and let E_N^* be the set of all other positive integers. Also, let

$$T_N = \prod_{p \in P[N]} \frac{1}{1 - a(p)} = \prod_{p \in P[N]} [1 + a(p) + a(p)^2 + \cdots].$$

As explained above, for each $n \in E_N$, the term $a(n)$ appears exactly once in this product, so, by (MS) (extended to finitely many series), we have $T_N = \sum_{n \in E_N} a(n)$. Now E_N certainly contains $1, 2, \ldots, N$, so

$$\left| \sum_{n=1}^{\infty} a(n) - T_N \right| = \left| \sum_{n \in E_N^*} a(n) \right| \leq \sum_{n > N} |a(n)|,$$

which tends to 0 as $N \to \infty$. □

Without the final step justifying convergence, the same reasoning also establishes the following:

Proposition 2.1.2 *Let F be a finite set of primes, and let E_F be the set of all numbers expressible as products of powers of members of F (including 1). Suppose that $a(n)$ is completely multiplicative and that $|a(p)| < 1$ for all $p \in F$. Then*

$$\sum_{n \in E_F} a(n) = \prod_{p \in F} \frac{1}{1 - a(p)}.$$

Proof As before. Note that the condition $|a(p)| < 1$ ensures absolute convergence of the geometric series. □

Our main example of 2.1.1 (and for the moment, our only one) is the following one, found by taking $a(n) = 1/n^s$:

Theorem 2.1.3 (the simple Euler product) *For* Re $s > 1$, *we have*

$$\zeta(s) = \prod_{p \in P} \left(1 - \frac{1}{p^s}\right)^{-1}. \qquad \square$$

At the same time, we have also proved the following statement, which was not at all obvious from the series expression.

Proposition 2.1.4 *We have* $\zeta(s) \neq 0$ *for* Re $s > 1$. $\square$

One can deduce further facts about the zeta function. For example, an argument of $\zeta(s)$ is given by $\sum_{p \in P} \arg(1 - p^{-s})$ if this converges: see exercise 3.

Divergence of $\sum_{p \in P}(1/p)$

We show next how, by a slight modification of the same reasoning, Euler proved that the series $\sum_{p \in P}(1/p)$ is divergent, at the same time obtaining an estimate for its partial sums. Recall from 1.4.2 that $\sum_{n=1}^{N} \frac{1}{n} > \log N$.

Lemma 2.1.5 *If* $0 < x < 1$, *then*

$$-\log(1-x) - x \leq \frac{x^2}{2(1-x)}.$$

Proof By the series for $\log(1-x)$,

$$-\log(1-x) - x = \sum_{n=2}^{\infty} \frac{x^n}{n} \leq \frac{1}{2} \sum_{n=2}^{\infty} x^n = \frac{x^2}{2(1-x)}. \qquad \square$$

Proposition 2.1.6 *The series* $\displaystyle\sum_{p \in P} \frac{1}{p}$ *is divergent. Further, for each* N,

$$\sum_{p \in P[N]} \frac{1}{p} \geq \log \log N - \tfrac{1}{2}.$$

Proof Define T_N as in the proof of 2.1.1, with $a(n) = 1/n$. Then

$$T_N = \prod_{p \in P[N]} \left(1 - \frac{1}{p}\right)^{-1} = \sum_{n \in E_N} \frac{1}{n} \geq \sum_{n=1}^{N} \frac{1}{n} > \log N.$$

Hence $\prod_{p \in P}(1 - 1/p)^{-1}$ is divergent. If $\sum_{p \in P}(1/p)$ were convergent, then so would be $\prod_{p\in P}(1-1/p)$ and $\prod_{p\in P}(1-1/p)^{-1}$. Hence $\sum_{p\in P}(1/p)$ is divergent. Furthermore, let

$$S_N = \sum_{p\in P[N]} \frac{1}{p}.$$

By 2.1.5, if $y > 1$, then

$$-\log\left(1-\frac{1}{y}\right) - \frac{1}{y} \leq \frac{1}{2y^2(1-1/y)} = \frac{1}{2y(y-1)}.$$

Hence

$$\begin{aligned}
\log T_N - S_N &= \sum_{p\in P[N]} \left[-\log\left(1-\frac{1}{p}\right) - \frac{1}{p}\right] \\
&\leq \frac{1}{2}\sum_{p\in P[N]} \frac{1}{p(p-1)} \\
&< \frac{1}{2}\sum_{n=2}^{\infty} \frac{1}{n(n-1)} \\
&= \frac{1}{2}\,.
\end{aligned}$$

So $S_N \geq \log T_N - \frac{1}{2} \geq \log\log N - \frac{1}{2}$. □

Note Recall that an opposite inequality was found in 1.6.5. We shall establish a more accurate estimation later (section 2.6): $S_N - \log\log N$ actually tends to a limit as $N \to \infty$. Although $\log\log N$ tends to infinity as $N \to \infty$, it does so *very* slowly! For example, $\log\log 10^{10} = \log(10\log 10) \approx 3.1366$.

As we have mentioned already, the Euler product opens up a whole vista of new ideas. Some of its applications are described in the next two sections of this chapter.

Exercises

1 Let A be the set of all integers of the form $2^r 3^s$. Evaluate

$$\sum_{n\in A}\frac{1}{n}, \qquad \sum_{n\in A}\frac{1}{n^2}.$$

2 *(An alternative proof of 2.1.6).* An integer n is *square-free* if there is no prime p such that $p^2|n$. Use prime factorization to show that

every integer is expressible as qm^2, where q is square-free. Let $F(N)$ be the set of square-free integers not greater than N. Show that

$$\sum_{n \le N} \frac{1}{n} \le \zeta(2) \sum_{q \in F(N)} \frac{1}{q}.$$

By considering $\prod_{p \in P[N]}(1 + 1/p)$ and using the fact that $1 + 1/p \le e^{1/p}$, prove that

$$\sum_{p \in P[N]} \frac{1}{p} \ge \log\log N - \log \zeta(2).$$

3 Let $0 < r < 1$. Draw a picture to illustrate the statement that the argument of $1 - re^{i\theta}$ is between $\pm \sin^{-1} r$ (but don't bother to prove it unless you want to). Use the Euler product to deduce that $\zeta(s) = Re^{i\alpha}$, where

$$|\alpha| \le \sum_{p \in P} \sin^{-1} \frac{1}{p^\sigma}.$$

Using the fact that $\sin^{-1}(x) \le (\pi/3)x$ for $0 \le x \le \frac{1}{2}$, deduce that Re $\zeta(s) > 0$ whenever $\sum_{p \in P}(1/p^\sigma) < \frac{3}{2}$. (*Note*: This occurs for $\sigma > 1.26$).

2.2 The Möbius function

Now consider the reciprocal of the Euler product. In the situation of Theorem 2.2.1, we have, clearly,

$$\frac{1}{\sum_{n=1}^{\infty} a(n)} = \prod_{p \in P} [1 - a(p)].$$

Let us consider this product. Write

$$Q_N = \prod_{p \in P[N]} [1 - a(p)].$$

If $n = p_1 p_2 \ldots p_k$, with each $p_j \le N$, then $a(n) = a(p_1) \ldots a(p_k)$, so $(-1)^k a(n)$ is a term in Q_N. However, if any p_j^2 is a factor of n, then there is no term $a(n)$. Also, $a(1) = 1$ is a term. Hence, if E_N (as before) is the set of integers expressible as products of primes in $P[N]$, then

$$Q_N = \sum_{n \in E_N} \mu(n) a(n),$$

where

$$\begin{aligned}
\mu(1) &= 1, \\
\mu(n) &= (-1)^k \quad \text{if } n = p_1 p_2 \dots p_k, \text{ a product of } k \text{ distinct primes}, \\
\mu(n) &= 0 \quad \text{if } p^2 | n \text{ for some prime } p.
\end{aligned}$$

This defines the *Möbius function* $\mu(n)$. A few of its values are

n	1	2	3	4	6	30
$\mu(n)$	1	-1	-1	0	1	-1

We now state the theorem suggested by the above discussion. Not much more is needed to complete the proof.

Theorem 2.2.1 *Let $a(n)$ be completely multiplicative, with $\sum_{n=1}^{\infty} |a(n)|$ convergent. Then*

$$\frac{1}{\sum_{n=1}^{\infty} a(n)} = \sum_{n=1}^{\infty} \mu(n) a(n).$$

Proof The series converges, since $|\mu(n)| \le 1$ for all n. Let its sum be S. With Q_N as above and $E_N^* = \mathbb{N} \setminus E_N$, we have

$$|S - Q_N| \le \sum_{n \in E_N^*} |a(n)| \le \sum_{n > N} |a(n)|,$$

which tends to 0 as $N \to \infty$. □

Note If F is a finite set of primes and E_F, $a(n)$ are as in 2.1.2, then clearly

$$\frac{1}{\sum_{n \in E_F} a(n)} = \prod_{p \in F} [1 - a(p)] = \sum_{n \in E_F} \mu(n) a(n).$$

The sum on the right is really a finite one, since it only includes those n that are products of members of F, without repetitions.

Of course, the particular case that is of most interest to us is

Theorem 2.2.2 *For* Re $s > 1$, *we have*

$$\frac{1}{\zeta(s)} = \sum_{n=1}^{\infty} \frac{\mu(n)}{n^s}.$$ □

We mention some immediate consequences of this identity. First, we write down the corresponding integral expression. Let $M(x) = \sum_{n \le x} \mu(n)$. Clearly, $M(x)$ is integer-valued and $|M(x)| \le x$ for all x (since $|\mu(n)| \le 1$). So, if Re $s > 1$, then $M(x)/x^s \to 0$ as $x \to \infty$, and hence we have immediately by 1.7.5:

Corollary 2.2.3 *Write* $M(x) = \sum_{n \le x} \mu(n)$. *Then, for* Re $s > 1$,

$$\frac{1}{\zeta(s)} = s \int_1^\infty \frac{M(x)}{x^{s+1}}\, dx. \qquad \square$$

Corollary 2.2.4 *For* $s = \sigma + it$ *with* $\sigma > 1$, *we have:*

$$\left| \frac{1}{\zeta(s)} \right| \le \zeta(\sigma), \qquad \left| \frac{1}{\zeta(s)} - 1 \right| \le \zeta(\sigma) - 1, \qquad |\zeta(s)| \ge \frac{\sigma - 1}{\sigma}.$$

Proof The first two statements follow at once from the series for $1/\zeta(s)$, since $|\mu(n)| \le 1$ for all n. The third statement follows from the fact (1.7.1) that $\zeta(\sigma) \le \sigma/(\sigma - 1)$. $\square$

Remark When $\zeta(\sigma) < 2$ (which occurs when $\sigma > 1.73$), the second inequality in 2.2.4 says that $\zeta(s)$ lies inside a certain circle (see exercise 9). Together with 1.7.4, this locates $\zeta(s)$ within the intersection of two circles.

Let us now give some more attention to the Möbius function itself. Given an arithmetic function $a(n)$, write (as in section 1.8), $F_a(s) = \sum_{n=1}^\infty a(n)/n^s$. With our previous notation u, e_1 for arithmetic functions, 2.2.2 translates the identity $\zeta(s)(1/\zeta(s)) = 1$ into

$$F_u(s) F_\mu(s) = F_{e_1}(s).$$

If we had a converse to 1.8.1, we could deduce that $u * \mu = e_1$. Such a converse does, in fact, exist, as a consequence of the uniqueness theorem for Dirichlet series (see section 1.7, exercise 9). However, instead of relying on this purely analytic theorem, we shall adopt the course of regarding the above relationship as simply "suggesting" that $u * \mu = e_1$, and going on to give a direct number-theoretic proof of this number-theoretic statement. Of course, once done, this gives a second proof of theorem 2.2.2.

Theorem 2.2.5 *We have* $u * \mu = e_1$ *(so* μ *is the inverse of* u *with respect to convolution). Equivalently, for* $n > 1$, *we have* $\sum_{i|n} \mu(i) = 0$.

Proof Clearly, $u(1)\mu(1) = 1$. Take $n > 1$. Recall that $(u*\mu)(n) = \sum_{i|n} \mu(i)$. Let n have prime factorization

$$n = p_1^{r_1} \dots p_k^{r_k}.$$

The $i > 1$ for which $\mu(i) \ne 0$ are of the form $i = q_1 \dots q_j$, where $q_1, \dots, q_j$ are a choice of j numbers from $p_1, \dots, p_k$. For a fixed $j \le k$, there are $\binom{k}{j}$

such choices, each defining a different i with $\mu(i) = (-1)^j$. Since $\mu(1) = 1$, this is also valid for $j = 0$ (note that $\binom{n}{0} = 1$), so, by the binomial theorem,

$$\sum_{i|n} \mu(i) = \sum_{j=0}^{k} \binom{k}{j} (-1)^j = (1-1)^k = 0. \qquad \square$$

Corollary 2.2.6 (Möbius inversion, first form) *For arithmetic functions a,b, the following statements are equivalent:*

(i) $b(n) = \sum_{i|n} a(i)$, *that is,* $b = a * u$,

(ii) $a(n) = \sum_{i|n} \mu(i)b(n/i)$, *that is,* $a = b * \mu$.

Proof If $b = a * u$, then

$$b * \mu = (a * u) * \mu = a * (u * \mu) = a * e_1 = a.$$

Conversely, if $a = b * \mu$, then $a * u = b * \mu * u = b * e_1 = b$. $\square$

Note As mentioned, 2.2.5 amounts to an alternative proof of 2.2.1. The proof of 2.2.5 is quite short, but it only *verifies* that μ (once we know about it) is the inverse of u, whereas the approach via the Euler product actually *led* us to the expression for μ, which was hardly obvious in advance!

Proposition 2.2.7 *If a is completely multiplicative, then the inverse of a with respect to convolution is* μa.

Proof Note that $a(1) = 1$, so that $ae_1 = e_1$. Applying 1.8.9 to the identity $u * \mu = e_1$, we obtain $(ua) * (\mu a) = ae_1$; in other words, $a * (\mu a) = e_1$. $\square$

We go on to derive a second form of Möbius inversion, which is suggested by summation of convolutions, as follows. Suppose that $f * u = g$, so that $g * \mu = f$. Write (as usual) $F(x) = \sum_{n \le x} f(n)$, and similarly $G(x)$. Then, by 1.8.4, $G(x) = \sum_{n \le x} F(x/n)$, and also $F(x) = \sum_{n \le x} \mu(n)G(x/n)$. In fact, this interaction between F and G holds even when they are not summation functions of arithmetic functions:

Proposition 2.2.8 (Möbius inversion, second form) *Let F be a function on* $(0, \infty)$ *that takes the value 0 on some interval* $(0, a)$. *For* $x > 0$, *let*

$$G(x) = \sum_{n=1}^{\infty} F\left(\frac{x}{n}\right).$$

Then

$$F(x) = \sum_{n=1}^{\infty} \mu(n) G\Big(\frac{x}{n}\Big).$$

Proof Note that $F(x/n) = 0$ for $n > x/a$, so the sum defining $G(x)$ is really a finite one, and there are no convergence problems. In the following, we write sums to infinity with this understanding. Clearly $G(x)$ is also zero for $0 < x < a$. Substituting $\mu * u = e_1$, we obtain

$$\begin{aligned} F(x) &= \sum_{j=1}^{\infty} e_1(j) F\Big(\frac{x}{j}\Big) \\ &= \sum_{j=1}^{\infty} F\Big(\frac{x}{j}\Big) \sum_{n|j} \mu(n). \end{aligned}$$

Reassemble this sum, first fixing n and then counting all multiples $j = kn$ of n. We obtain

$$\begin{aligned} F(x) &= \sum_{n=1}^{\infty} \mu(n) \sum_{k=1}^{\infty} F\Big(\frac{x}{kn}\Big) \\ &= \sum_{n=1}^{\infty} \mu(n) G\Big(\frac{x}{n}\Big). \end{aligned}$$

□

From 2.2.2 and the fact that $\zeta(\sigma) \to \infty$ as $\sigma \to 1^+$, one might expect that the series $\sum_{n=1}^{\infty} \mu(n)/n$ converges to 0. This fairly harmless sounding statement turns out to be a theorem in the same family as the prime number theorem itself, and we will eventually prove it in chapter 3. However, it is quite easy to show that the partial sums of the series are bounded, and we do this now. The proof is a beautiful example of the estimation of a quantity by considering the summation function of a suitable convolution.

Proposition 2.2.9 *For all positive integers N, we have*

$$\left| \sum_{n=1}^{N} \frac{\mu(n)}{n} \right| \leq 1.$$

Proof By the identity $\mu * u = e_1$ and 1.8.5, we have

$$1 = \sum_{n=1}^{N} (\mu * u)(n) = \sum_{n=1}^{N} \mu(n) \left[\frac{N}{n} \right].$$

This equates to

$$\sum_{n=1}^{N} \mu(n) \left(\frac{N}{n} - \left\{ \frac{N}{n} \right\} \right) = N \sum_{n=1}^{N} \frac{\mu(n)}{n} - r_N,$$

where $r_N = \sum_{n=1}^{N} \mu(n)\{N/n\}$. Now $\{N/1\} = \{N\} = 0$, so

$$|r_N| = \left| \sum_{n=2}^{N} \mu(n) \left\{ \frac{N}{n} \right\} \right| \leq \sum_{n=2}^{N} |\mu(n)| \leq N - 1.$$

Hence

$$N \left| \sum_{n=1}^{N} \frac{\mu(n)}{n} \right| \leq |r_N| + 1 \leq N,$$

which gives the statement. □

Further series derived from the Euler product

This subsection, though (in the author's view!) interesting, is not essential for the proof of the prime number theorem. Let $a(n)$ be as in 2.1.1. Write

$$S_1 = \sum_{n=1}^{\infty} a(n), \qquad S_2 = \sum_{n=1}^{\infty} a(n)^2.$$

Then, by 2.1.1,

$$\frac{S_1}{S_2} = \prod_{p \in P} \frac{1 - a(p)^2}{1 - a(p)} = \prod_{p \in P} [1 + a(p)].$$

Reasoning exactly as in 2.2.1, without the -1 factors, we deduce:

Proposition 2.2.10 *Let $a(n)$ be completely multiplicative, with $\sum_{n=1}^{\infty} |a(n)|$ convergent. Then*

$$\frac{S_1}{S_2} = \sum_{n=1}^{\infty} |\mu(n)| a(n).$$

In particular, for Re $s > 1$,

$$\frac{\zeta(s)}{\zeta(2s)} = \sum_{n=1}^{\infty} \frac{|\mu(n)|}{n^s}.$$ □

Corollary 2.2.11 (strengthening 2.2.4) *For $s = \sigma + it$ with $\sigma > 1$,*

$$\left| \frac{1}{\zeta(s)} \right| \leq \frac{\zeta(\sigma)}{\zeta(2\sigma)} \quad \text{and} \quad \left| \frac{1}{\zeta(s)} - 1 \right| \leq \frac{\zeta(\sigma)}{\zeta(2\sigma)} - 1.$$ □

Note that $|\mu(n)|$ can also be written as $\mu(n)^2$. It can be described very simply: it takes the value 1 when n is *square-free*, that is, when there is no prime p such that $p^2|n$.

Recall that *Liouville's function* is defined by $\lambda(n) = (-1)^{\Omega(n)}$, where $\Omega(n)$ is the number of prime factors of n, counted with repetitions (also $\lambda(1) = 1$). It is completely multiplicative, and clearly $\lambda(p) = -1$ for prime p.

Proposition 2.2.12 *Let $a(n)$ be as in 2.2.10. Then, with the above notation,*

$$\frac{S_2}{S_1} = \sum_{n=1}^{\infty} \lambda(n)a(n).$$

In particular, for $\operatorname{Re} s > 1$,

$$\frac{\zeta(2s)}{\zeta(s)} = \sum_{n=1}^{\infty} \frac{\lambda(n)}{n^s}.$$

Proof As shown above, $S_2/S_1 = \prod_{p \in P}[1 + a(p)]^{-1}$. Let $b(n) = \lambda(n)a(n)$. Then b is completely multiplicative and $b(p) = -a(p)$, so directly from the generalized Euler product (2.1.1), we have $S_2/S_1 = \sum_{n=1}^{\infty} b(n)$. □

We finish with a direct proof of the convolution identity suggested by the equality

$$\frac{\zeta(2s)}{\zeta(s)}\zeta(s) = \zeta(2s).$$

Recall that $\zeta(2s) = \sum_{n=1}^{\infty} 1/n^{2s} = \sum_{n=1}^{\infty} u_S(n)/n^s$, where S is the set of squares.

Proposition 2.2.13 *We have* $\lambda * u = u_S$.

Proof By 1.8.8, $\lambda * u$ is multiplicative. So the statement follows if we can show that, if p is prime, then $(\lambda * u)(p^n)$ equals 1 when n is even and 0 when n is odd. But this is true, because

$$(\lambda * u)(p^n) = \sum_{r=0}^{n} \lambda(p^r) = \sum_{r=0}^{n} (-1)^r. \qquad \square$$

As well as *suggesting* convolution identities like this one, the corresponding series provide a helpful way to remember (or reconstruct) them.

Euler's function ϕ

This is another of the basic functions of number theory (though not, as it happens, of much importance for the prime number theorem). It is defined as follows (for $n \geq 1$): let E_n be the set of integers r such that $1 \leq r \leq n$ and $(r, n) = 1$. Then $\phi(n)$ is the number of members of E_n.

Clearly, $\phi(n) \leq n-1$ for $n \geq 2$, and if p is prime, then $\phi(p) = p-1$. Also, $\phi(1) = 1$.

Example The members of E_{30} are 1, 7, 11, 13, 17, 19, 23, 29, so $\phi(30) = 8$. Note that $30 = 2 \times 3 \times 5$, and these are the numbers below 30 that are not multiples of 2, 3 or 5.

We shall derive, rather briefly, the well-known formula for $\phi(n)$ and describe its connection to the Möbius function.

Lemma 2.2.14 *The function ϕ is multiplicative.*

Proof This is a consequence of the "Chinese remainder theorem". Let $(m, n) = 1$. By elementary number theory, we have $(x, mn) = 1$ if and only if $(x, m) = (x, n) = 1$. In turn, this is equivalent to the pair of statements:

$$x \equiv a \pmod{m}, \quad \text{where } a \in E_m,$$

$$x \equiv b \pmod{n}, \quad \text{where } b \in E_n.$$

By the Chinese remainder theorem, for each pair (a, b) in $E_m \times E_n$, there is a unique x such that $1 \leq x < mn$ and this pair of statements holds. So the number of such x is $\phi(m)\phi(n)$. □

Proposition 2.2.15 *If the prime factorization of n is $\prod_{j=1}^{m} p_j^{k_j}$, then*

$$\phi(n) = \prod_{j=1}^{m} p_j^{k_j - 1}(p_j - 1) = n \prod_{j=1}^{m} \left(1 - \frac{1}{p_j}\right).$$

Proof The statement follows by the lemma if we can show that $\phi(p^k) = p^{k-1}(p-1)$ when p is prime. The integers $r \leq p^k$ with $(r, p^k) > 1$ are the multiples of p, in other words, jp for $1 \leq j \leq p^{k-1}$. The number of these is p^{k-1}, so $\phi(p^k) = p^k - p^{k-1}$, as required. □

Proposition 2.2.16 *For all $n \geq 1$,*

$$\phi(n) = n \sum_{k|n} \frac{\mu(k)}{k}.$$

In convolution form, $\phi = \mu * I$, where $I(n) = n$.

Proof By the remark after 2.2.1,

$$\prod_{j=1}^{m}\left(1-\frac{1}{p_j}\right)=\sum_{k\in E_F}\frac{\mu(k)}{k},$$

where E_F is the set of products formed from the numbers p_j without repetition. Clearly, E_F comprises the set of divisors k of n having $\mu(k)\neq 0$. The stated equality follows, by 2.2.15. The convolution identity is simply a restatement. □

Corollary 2.2.17 *We have* $\phi * u = I$. *Equivalently,* $\sum_{j|n}\phi(j)=n$. □

Alternatively, one can prove 2.2.17 directly and work back to 2.2.15. The identity $\phi=\mu * I$ gives at once

$$\sum_{n=1}^{\infty}\frac{\phi(n)}{n^s}=\frac{\zeta(s-1)}{\zeta(s)}\qquad\text{for Re } s>2.$$

Exercises

1 Tabulate $\mu(n)$ and $M(n)=\sum_{j\le n}\mu(j)$ for $1\le n\le 40$.

2 Show that μ is multiplicative (consider cases, e.g., $\mu(m)$ and $\mu(n)$ both non-zero).

3 Show that if a is multiplicative and $p_1,\ldots,p_k$ are the prime divisors of n, then

$$\prod_{j=1}^{k}(1-a(p_j))=((\mu a)*u)(n).$$

4 Formulate and prove the convolution identity suggested by the Dirichlet series for $\zeta(s)/\zeta(2s)$ and $\zeta(2s)/\zeta(s)$.

5 Repeat exercise 4, considering the series for $\zeta(s)/\zeta(2s)$ and $\zeta(2s)$.

6 Let λ be the Liouville function. Show that $\left|\sum_{n=1}^{N}\lambda(n)/n\right|\le\frac{3}{2}$ for all $N\ge 1$.

7 Copy the method of 1.4.5 to prove that, for integers n,

$$\sum_{j\le n}\frac{\phi(j)}{j}\ge\tfrac{1}{2}(n+1).$$

8 Show that there are $2^{\omega(n)}$ divisors j of n that have $\mu(j)^2=1$. Deduce that $(\mu^2 * u)(n)=2^{\omega(n)}$ for all n, and hence express $\sum_{n=1}^{\infty}2^{\omega(n)}/n^s$ in terms of the zeta function.

9 Let ρ be a real number between 1 and 2. Show that the complex numbers z satisfying $|1/z - 1| \le \rho - 1$ form the inside of a circle, and find the two points where this circle meets the real axis. Deduce that for all $s = \sigma + it$ such that $\zeta(\sigma) < 2\zeta(2\sigma)$ we have $\operatorname{Re} \zeta(s) \ge \zeta(2\sigma)/\zeta(\sigma)$ $(> \frac{1}{2})$. (*Remark*: This condition is satisfied when $\sigma > 1.6$).

2.3 The series for $\log \zeta(s)$ and $\zeta'(s)/\zeta(s)$

The series for $\log \zeta(s)$

In this section, we continue to exploit the Euler product for $\zeta(s)$, deriving Dirichlet series for the functions in the title. Both these series are essential ingredients of the proof of the prime number theorem.

First, consider $\log \zeta(s)$. A *logarithm* of a complex number z is any w such that $e^w = z$. As the reader will know, there are many such w's. If complex numbers z_j $(j \in \mathbb{N})$ are given, and logarithms w_j of z_j can be chosen so that $\sum_{j=1}^{\infty} w_j$ is convergent (say to w), then, by continuity, $e^w = \prod_{j=1}^{\infty} z_j$, so w is a logarithm of $\prod_{j=1}^{\infty} z_j$. So, by the Euler product, we will obtain a logarithm of $\zeta(s)$ if we can choose (for each p) a logarithm w_p of $(1-p^{-s})^{-1}$ in such a way that $\sum_{p \in P} w_p$ converges. We now show that this is achieved by using the familiar logarithmic series, and furthermore that the resulting function is holomorphic (alias differentiable).

Lemma 2.3.1 *Define* $h(z) = \sum_{m=1}^{\infty} (z^m/m)$ *for* $|z| < 1$. *Then* $h(z)$ *is a logarithm of* $1/(1-z)$. *Also,* $|h(z)| \le 2|z|$ *when* $|z| \le \frac{1}{2}$.

Proof The series for $h(z)$ certainly converges when $|z| < 1$. By termwise differentiation (which is valid for power series), we have

$$h'(z) = \sum_{m=1}^{\infty} z^{m-1} = \frac{1}{1-z},$$

hence

$$\frac{d}{dz}(1-z)e^{h(z)} = -e^{h(z)} + \frac{1-z}{1-z}e^{h(z)} = 0.$$

Hence $(1-z)e^{h(z)}$ is constant (say $= c$) for $|z| < 1$. Take $z = 0$ to get $c = e^0 = 1$. Further, when $|z| \le \frac{1}{2}$,

$$|h(z)| \le \sum_{m=1}^{\infty} |z|^m = \frac{|z|}{1-|z|} \le 2|z|. \qquad \square$$

We now use the following theorem on "double series" (for a proof, see appendix B):

(DS) *Suppose that the repeated sum* $\sum_{j=1}^{\infty}\sum_{k=1}^{\infty}|a_{j,k}|$ *converges. Then the repeated sums*

$$\sum_{j=1}^{\infty}\sum_{k=1}^{\infty} a_{j,k}, \qquad \sum_{k=1}^{\infty}\sum_{j=1}^{\infty} a_{j,k}$$

both converge to the same sum, say S. Further, if $\sum_{n=1}^{\infty} c_n$ *is any series obtained by arranging the terms* $a_{j,k}$ *as a single series, then* $\sum_{n=1}^{\infty} c_n$ *converges to S.*

Theorem 2.3.2 *For* Re $s > 1$, *a holomorphic function, forming a logarithm of* $\zeta(s)$, *is defined by*

$$H(s) = \sum_{p\in P}\sum_{m=1}^{\infty} \frac{1}{mp^{ms}} = \sum_{n=1}^{\infty} \frac{c(n)}{n^s},$$

where

$$c(n) = \begin{cases} 1/m & \text{if } n = p^m \text{ for some prime } p \text{ and some integer } m \\ 0 & \text{otherwise.} \end{cases}$$

Both series are absolutely convergent.

Proof Provided that the series $\sum_{p\in P} h(1/p^s)$ converges, its sum will be a logarithm of $\zeta(s)$. This sum is the repeated sum

$$\sum_{p\in P}\sum_{m=1}^{\infty} \frac{1}{mp^{ms}}.$$

Since $|p^{ms}| = p^{m\sigma}$, the required convergence, as well as the stated rearrangement as a single series, will follow if we know that

$$\sum_{p\in P}\sum_{m=1}^{\infty} \frac{1}{mp^{m\sigma}}$$

is convergent. But this is true, since

$$\sum_{m=1}^{\infty} \frac{1}{mp^{m\sigma}} = h\left(\frac{1}{p^\sigma}\right) \le \frac{2}{p^\sigma}$$

and $\sum_{p\in P}(2/p^\sigma)$ is convergent. By 1.7.10, the single series defines a holomorphic function. □

We shall use the notation $\log\zeta(s)$ (for $\mathrm{Re}\, s > 1$) to mean the function $H(s)$ just defined. It does not necessarily coincide with the so-called "principal value" of the logarithm with argument between $-\pi$ and π. However, it is clearly the real-valued logarithm when z is real.

The series $\sum_{p\in P}(1/p^s)$

The double series for $\log\zeta(s)$ sheds some light on the series $\sum_{p\in P}(1/p^s)$, in which the prime terms have been picked out of the series for $\zeta(s)$. In the double series, we can reverse the order of summation (again using (DS)) and separate out the term $m = 1$, as follows:

$$\begin{aligned}\log\zeta(s) &= \sum_{m=1}^{\infty}\frac{1}{m}\sum_{p\in P}\frac{1}{p^{ms}}\\ &= \sum_{p\in P}\frac{1}{p^s} + \sum_{m=2}^{\infty}\frac{1}{m}\sum_{p\in P}\frac{1}{p^{ms}}.\end{aligned}$$

The point is that the second sum is uniformly quite small. The next result gives the details.

Proposition 2.3.3 *For real $\sigma > 1$, we have*

$$\sum_{p\in P}\frac{1}{p^\sigma} < \log\zeta(\sigma).$$

For all complex s with $\mathrm{Re}\, s > 1$, *we have*

$$\left|\sum_{p\in P}\frac{1}{p^s} - \log\zeta(s)\right| < \frac{1}{2}.$$

Proof The first statement follows immediately from the double series as written above. To prove the second statement, we have, for all $\sigma > 1$,

$$\begin{aligned}\left|\log\zeta(s) - \sum_{p\in P}\frac{1}{p^s}\right| &\le \sum_{m=2}^{\infty}\frac{1}{m}\sum_{p\in P}\frac{1}{p^{m\sigma}}\\ &\le \frac{1}{2}\sum_{p\in P}\sum_{m=2}^{\infty}\frac{1}{p^m}\\ &= \frac{1}{2}\sum_{p\in P}\frac{1}{p(p-1)}\end{aligned}$$

$$\leq \frac{1}{2}\sum_{n=2}^{\infty}\frac{1}{n(n-1)}$$
$$= \frac{1}{2}. \qquad \square$$

Hence, in particular, $\sum_{p\in P}(1/p^\sigma) \to \infty$ as $\sigma \to 1^+$. This fact is a natural companion to the divergence of $\sum_{p\in P}(1/p)$, which we proved in a similar way in 2.1.6.

The series for $\zeta'(s)/\zeta(s)$

We claim to have selected a logarithm of $\zeta(s)$ that possesses a derivative. Let us now consider this derivative. Recall that for a real function $f(x)$, the derivative of $\log f(x)$ is $f'(x)/f(x)$. The same applies to complex functions in the following sense. If $g(s)$ is differentiable and $e^{g(s)} = f(s)$, then differentiation gives $g'(s)e^{g(s)} = f'(s)$, so that $g'(s) = f'(s)/f(s)$.

In our case, with $H(s)$ defined as above, this means that

$$\frac{\zeta'(s)}{\zeta(s)} = H'(s)$$
$$= -\sum_{n=1}^{\infty}\frac{c(n)\log n}{n^s} \qquad \text{by 1.7.10.}$$

Now, for $n = p^m$,

$$c(n)\log n = \frac{1}{m}m\log p = \log p.$$

In other words, the following is true:

Theorem 2.3.4 *For* Re $s > 1$, *we have*

$$\frac{\zeta'(s)}{\zeta(s)} = -\sum_{n=1}^{\infty}\frac{\Lambda(n)}{n^s},$$

where

$$\Lambda(n) = \begin{cases} \log p & \textit{if } n = p^m \textit{ for some prime } p \textit{ and some integer } m, \\ 0 & \textit{otherwise.} \end{cases} \qquad \square$$

The function $\Lambda(n)$ is called the *von Mangoldt* function (and $\Lambda(n)$, again for historical reasons, is the standard notation). Exactly as with the Möbius function, the Euler product has again led us to an arithmetic function that would hardly have been obvious to guess!

Now that $\Lambda(n)$ has been identified for us, let us formulate and verify the convolution identity corresponding to this series. The series identity

$$\frac{\zeta'(s)}{\zeta(s)}\zeta(s) = \zeta'(s)$$

suggests the convolution identity $\Lambda * u = \ell$, where $\ell(n) = \log n$. As before, we shall now prove this identity directly, thereby giving a second proof of Theorem 2.3.4 (note that convergence of the series for Re $s > 1$ follows from the fact that $\Lambda(n) \leq \log n$ for all n). At the same time, the Möbius function gives a second, equivalent identity.

Theorem 2.3.5 *Write $\ell(n) = \log n$. Then*

$$\Lambda * u = \ell \quad \text{and} \quad \ell * \mu = \Lambda.$$

Proof The second statement follows from the first one, since $u * \mu = e_1$. To prove the first statement, note first that $\Lambda(1) = \ell(1) = 0$. Now choose $n > 1$, with prime factorization

$$n = p_1^{r_1} \dots p_k^{r_k}.$$

The divisors i of n for which $\Lambda(i) \neq 0$ are the numbers $p_j^{s_j}$ for $1 \leq j \leq k$ and $1 \leq s_j \leq r_j$. Each of these has $\Lambda(i) = \log p_j$. Hence

$$\begin{aligned}(\Lambda * u)(n) &= \sum_{i|n} \Lambda(i) \\ &= \sum_{j=1}^{k} r_j \log p_j \\ &= \log n.\end{aligned}$$

□

The series for $\zeta'(s)/\zeta(s)$ plays an absolutely central part in the strategy to prove the prime number theorem. At this point, it is nearly, but not quite, clear why. Of course, our goal is to prove that $\theta(x) \sim x$. The general idea is that by knowing enough about a function defined by a Dirichlet series, we hope to derive information about the coefficients of the series – more exactly, about their partial sums. The Dirichlet series whose coefficients would deliver $\theta(x)$ in this way is, of course, the derivative of $\sum_{p \in P} 1/p^s$. However, this is an awkward function, and it is easier to establish properties of $\zeta'(s)/\zeta(s)$. The summation function of its coefficients $\Lambda(n)$ differs from $\theta(x)$ by including powers of primes as well as primes. The next task is to show that this summation function does just as well as $\theta(x)$ for our purpose. This is the subject of the next section.

Exercises

1 What convolution identity is suggested by the fact that $\zeta(1/\zeta)' = -\zeta'/\zeta$? Prove this identity by showing that $\ell(u*\mu) = 0$ and applying a result from section 1.8.

2 Let

$$b(n) = \begin{cases} \log n & \text{if } n \text{ is prime}, \\ 0 & \text{otherwise}. \end{cases}$$

Describe $(b*u)(n)$ in terms of the prime factorization of n.

3 Let $f(s) = -\zeta'(s)/\zeta(s)$. Show that $|f(s)| \leq f(\sigma)|$ for $s = \sigma + it$ with $\sigma > 1$. Using estimations from section 1.7, deduce that, for such s,

$$|\zeta'(s)| \leq \frac{\sigma}{\sigma - 1}|\zeta(s)|.$$

4 Use Abel summation to show that

$$\log \zeta(s) = s \int_2^\infty \frac{\pi(x)}{x(x^s - 1)}\, dx.$$

5 By termwise differentiation of the series $\sum_{p \in P} h(1/p^s)$, show that

$$-\frac{\zeta'(s)}{\zeta(s)} = \sum_{p \in P} \frac{\log p}{p^s - 1},$$

and express this as an integral involving $\theta(x)$.

2.4 Chebyshev's psi function and powers of primes

Estimation of $\psi(x)$ and comparison with $\theta(x)$

As explained at the end of the previous section, the series $-\zeta'(s)/\zeta(s) = \sum_{n=1}^\infty \Lambda(n)/n^s$ leads us to consider, as an alternative to Chebyshev's $\theta(x)$, the summation function of $\Lambda(n)$. Chebyshev also introduced this function, and his notation for it, $\psi(x)$, remains in use:

$$\psi(x) = \sum_{n \leq x} \Lambda(n).$$

Another way to describe it is as follows. As before, let $p_1, p_2, \ldots, p_n$ be the primes not greater than x. For each $j \leq n$, let k_j be the largest k such that $p_j^k \leq x$. Then each p_j^k, for $1 \leq k \leq k_j$, contributes $\log p_j$ to $\psi(x)$, and hence

$$\psi(x) = k_1 \log p_1 + \cdots + k_n \log p_n.$$

A pleasantly easy deduction is the following inequality.

Proposition 2.4.1 *For all $x > 0$, we have $\psi(x) \leq \pi(x) \log x$.*

Proof With the above notation, $n = \pi(x)$ and $k_j \log p_j \leq \log x$ for each $j \leq n$. □

Hence certainly $\psi(x) < x \log x$, and in fact, by 1.6.4, $\psi(x) \leq Cx$ for some constant C (we return below to the question of estimating this C). So, if Re $s > 1$, then $\psi(x)/x^s \to 0$ as $x \to \infty$. This is the condition needed to rewrite the series $\sum_{n=1}^{\infty} \Lambda(n)/n^s$ in integral form (see 1.7.5). We obtain:

Proposition 2.4.2 *For* Re $s > 1$, *we have*

$$\frac{\zeta'(s)}{\zeta(s)} = -s \int_1^{\infty} \frac{\psi(x)}{x^{s+1}}\, dx.$$ □

Now let us compare $\psi(x)$ with $\theta(x)$. Obviously, $\psi(x) \geq \theta(x)$. The exact relationship can be expressed as follows.

Proposition 2.4.3 *Let m be the largest integer such that $2^m \leq x$. Then*

$$\psi(x) = \theta(x) + \theta(x^{1/2}) + \theta(x^{1/3}) + \cdots + \theta(x^{1/m}).$$

Proof For each $k \geq 2$, the difference $\psi(x) - \theta(x)$ contains a term $\log p$ for each prime p with $p^k \leq x$, or $p \leq x^{1/k}$. There only are any such primes if $k \leq m$. For fixed k, the sum of these terms is, by definition, $\theta(x^{1/k})$. The statement follows. □

Using Chebyshev's estimate $\theta(x) \leq (\log 4)x$, we deduce:

Proposition 2.4.4 *For all $x > 1$, we have $\psi(x) - \theta(x) \leq 6x^{1/2}$. Also, given $\varepsilon > 0$, we have $\psi(x) - \theta(x) \leq (\log 4 + \varepsilon)x$ for all large enough x.*

Proof By the expression in 2.4.3,

$$\psi(x) - \theta(x) \leq \theta(x^{1/2}) + m\theta(x^{1/3}),$$

where $m \leq (\log x / \log 2)$. So, by Chebyshev's estimate,

$$\psi(x) - \theta(x) \leq x^{1/2} \log 4 + 2x^{1/3} \log x.$$

Clearly, the second term is not greater than $\varepsilon x^{1/2}$ once x is large enough. Also, one finds by differentiation that the greatest value of $\log x / x^{\alpha}$ is $1/(\alpha e)$. Hence $x^{1/3} \log x \leq 6e^{-1} x^{1/2}$, and

$$\psi(x) - \theta(x) \leq (\log 4 + 12e^{-1}) x^{1/2} < 6x^{1/2}.$$ □

With more effort, one could replace the 6 by a smaller constant. An immediate consequence of 2.4.4 is the following corollary, which shows that $\psi(x)$ will do as well as $\theta(x)$ for the prime number theorem.

Corollary 2.4.5 *We have*

$$\frac{1}{x}[\psi(x) - \theta(x)] \to 0 \qquad \text{as } x \to \infty.$$

Hence if one of $\theta(x)/x$, $\psi(x)/x$ tends to a limit L as $x \to \infty$, then so does the other. □

Note One can easily prove this corollary using only the trivial inequality $\theta(x) \le x \log x$ instead of Chebyshev's estimate. For, inserted into the proof of 2.4.4, this gives (with $c = 1/\log 2$):

$$\begin{aligned} \psi(x) - \theta(x) &\le \tfrac{1}{2}x^{1/2} \log x + c \log x . \tfrac{1}{3}x^{1/3} \log x \\ &\le (\tfrac{1}{2} + 2ce^{-1})x^{1/2} \log x \\ &< 2x^{1/2} \log x. \end{aligned}$$

We now give a more precise version of Chebyshev's upper estimate adapted to $\psi(x)$ instead of $\theta(x)$.

Theorem 2.4.6 *For all $x > 1$, we have $\psi(x) < 2x$. Also, given $\varepsilon > 0$, we have $\psi(x) \le (\log 4 + \varepsilon)x$ for all large enough x.*

Proof By 2.4.4,

$$\frac{\psi(x)}{x} \le \frac{\theta(x)}{x} + \frac{6}{x^{1/2}} \le \log 4 + \frac{6}{x^{1/2}}.$$

Clearly, this is less than $\log 4 + \varepsilon$ once x is large enough. Also, it is less than $1.4 + 0.6 = 2$ when $x \ge 100$. To dispose of numbers below 100, note first that

$$\psi(100) - \theta(100) = 5 \log 2 + 3 \log 3 + \log 5 + \log 7 \approx 10.317.$$

This is enough to show that

$$\frac{\psi(x)}{x} \le \log 4 + \frac{1}{x}[\psi(x) - \theta(x)] < 2 \qquad \text{for } 17 \le x \le 100.$$

Finally, one checks easily that $\psi(n) < 2n$ (even, in fact, $\psi(n) < n$) for each integer $n \le 16$. □

A numerical example. The following table gives values of $\psi(x)$ in the same range $181 \le x < 211$ considered for $\theta(x)$ in section 1.6.

x	$[181, 191)$	$[191, 193)$	$[193, 197)$	$[197, 199)$	$[199, 211)$
$\psi(x)$	185.05	190.30	195.57	200.85	206.14

We see that $\psi(x) - x$ changes sign five times in this range. The difference $\psi(x) - \theta(x)$ remains constant at 17.58 throughout the range.

Powers of primes

The move from $\theta(x)$ to $\psi(x)$ amounted to including powers of primes as well as primes themselves. It is of interest to consider the frequency of such numbers, without the weighting factor $\log p$. Would it make much difference if they were included with primes in the prime number theorem? At first, one forms the impression that there are quite a lot of them: $4, 8, 9, 16, 25, 27, 32, \ldots$. Write $\pi^*(x)$ for the number of integers not greater than x that are of the form p^k for some prime p and some integer k. We can estimate $\pi^*(x)$ by methods similar to those we have just been using.

Proposition 2.4.7 *We have*

$$\pi^*(x) = \pi(x) + \pi(x^{1/2}) + \cdots + \pi(x^{1/m}),$$

where m is the largest integer such that $2^m \le x$. If C is such that $\pi(x) \le Cx/\log x$ for all $x \ge 2$, then

$$\pi^*(x) - \pi(x) \le 12\, C\frac{x^{1/2}}{\log x} \qquad \textit{for all } x \ge 2.$$

Proof For fixed k, the number of primes p with $p^k \le x$ is $\pi(x^{1/k})$: hence the given expression. Using the inequality $x^{1/3}\log x \le 6e^{-1}x^{1/2}$ as in the proof of 2.4.4, we find

$$\begin{aligned}
\pi^*(x) - \pi(x) &\le \pi(x^{1/2}) + m\pi(x^{1/3}) \\
&\le \frac{2Cx^{1/2}}{\log x} + \frac{\log x}{\log 2}\,\frac{3Cx^{1/3}}{\log x} \\
&\le \frac{Cx^{1/2}}{\log x}\left(2 + \frac{18}{e\log 2}\right) \\
&\le 12\, C\frac{x^{1/2}}{\log x}. \qquad \square
\end{aligned}$$

The message is that powers of primes are really very sparse! A numerical example will underline this point. From the expression in 2.4.7, we find that $\pi^*(10^6) - \pi(10^6) = 236$ (while $\pi(10^6)$ itself is 78,498). Of course, this evaluation only requires values of $\pi(x)$ for $x \leq 1,000$.

One way of regarding powers of primes is that they are defined by the property $\omega(n) = 1$, while primes are defined by $\Omega(n) = 1$.

Chebyshev's lower estimate

With the help of convolutions, we can now give a rather elegant proof of this estimate, as promised in section 1.6. It is obtained in the first place for $\psi(x)$ rather than $\theta(x)$.

Let ν be the arithmetic function $e_1 - 2e_2$, so that $\nu(1) = 1$, $\nu(2) = -2$ and $\nu(j) = 0$ for $j > 2$. Clearly,

$$(u * \nu)(n) = \begin{cases} 1 & \text{if } n \text{ is odd,} \\ -1 & \text{if } n \text{ is even.} \end{cases}$$

Let $E(x) = \sum_{n \leq x} (u * \nu)(n)$. Then

$$E(x) = \begin{cases} 1 & \text{if } [x] \text{ is odd,} \\ 0 & \text{if } [x] \text{ is even.} \end{cases}$$

Lemma 2.4.8 *For any $x \geq 2$,*

$$\sum_{j \leq x} \Lambda(j) E\left(\frac{x}{j}\right) = \sum_{k \leq x} \log k - 2 \sum_{k \leq x/2} \log k.$$

Proof By 1.8.4 and the identity $\Lambda * u = \ell$, we have

$$\begin{aligned} \sum_{j \leq x} \Lambda(j) E\left(\frac{x}{j}\right) &= \sum_{j \leq x} [\Lambda * (u * \nu)](j) \\ &= \sum_{j \leq x} (\ell * \nu)(j) \\ &= \sum_{j \leq x} \nu(j) \sum_{k \leq x/j} \log k \qquad \text{by 1.8.4 again} \\ &= \sum_{k \leq x} \log k - 2 \sum_{k \leq x/2} \log k. \end{aligned}$$

□

Lemma 2.4.9 *For positive integers n, we have* $\psi(2n) \geq \log \binom{2n}{n}$.

Proof Let S be the sum in the previous lemma, with $x = 2n$. Then $S \le \psi(2n)$, since $E(2n/j) \le 1$ for all j. Meanwhile, we have

$$\sum_{k \le 2n} \log k - 2\sum_{k \le n} \log k = \sum_{k=n+1}^{2n} \log k - \sum_{k=1}^{n} \log k = \log \binom{2n}{n}. \qquad \square$$

Proposition 2.4.10 *Let $\varepsilon > 0$. Then, for sufficiently large x:*

$$\psi(x) \ge (\log 2 - \varepsilon)x,$$

$$\theta(x) \ge (\log 2 - \varepsilon)x,$$

$$\pi(x) \ge (\log 2 - \varepsilon)\,\mathrm{li}(x).$$

Proof Let $N = \binom{2n}{n}$. It is easily verified that N is the largest of the $2n+1$ terms in the binomial expansion of $(1+1)^{2n}$, so we have $2^{2n} \le (2n+1)N$. So, by 2.4.9,

$$\psi(2n) \ge \log N \ge 2n \log 2 - \log(2n+1).$$

Given x, let n be such that $2n \le x < 2n+2$. Then

$$\psi(x) \ge (x-2)\log 2 - \log(x+1),$$

from which it is clear that $\psi(x)/x \ge \log 2 - \varepsilon$ for sufficiently large x. The statement for $\theta(x)$ follows by 2.4.5 and the statment for $\pi(x)$ by 1.6.2. $\square$

Clearly this gives an opposite inequality to 2.4.4: there is a constant $c > 0$ such that $\psi(x) - \theta(x) \ge cx^{1/2}$ for large enough x.

Note 1 The weaker statement $\pi(x) \ge (\log 2 - \varepsilon)x/\log x$ (for large enough x) follows at once from the statement for $\psi(x)$, by the trivial inequality $\psi(x) \le \pi(x)\log x$ (see 2.4.1), without 1.6.2.

Note 2 Of course, there is some positive constant c such that $\psi(x) \ge cx$ for all $x \ge 2$, since such an estimate certainly applies on any bounded interval $[2, x_0]$. A similar comment applies to each of the other lower estimates. We will not spend any time on finding actual values for these constants.

Note 3 Since $E(x/n) = 1$ for $\frac{1}{2}x < n \le x$, we also have $S \ge \psi(x) - \psi(x/2)$. This can be used to derive Chebyshev's upper estimate, though not quite in the form $\theta(x) \le (\log 4)x$.

Exercises

1 Show that

$$\sum_{n\leq x}\psi\left(\frac{x}{n}\right)=\sum_{n\leq x}\log n.$$

2 Use Chebyshev's lower estimate to show that, for some constant c,

$$\sum_{p\in P[x]}\frac{1}{p}\geq c\log\log x$$

for all $x\geq 2$.

3 Show that $\psi(n)$ is the logarithm of the lowest common multiple of $2,3,\ldots,n$.

4 Use Möbius inversion to show that

$$\theta(x)=\sum_{n=1}^{m}\mu(n)\psi(x^{1/n}),$$

where m is the largest integer such that $2^m\leq x$. (Write $F(t)=\theta(e^t)$.)

5 Let $c(n)=\Lambda(n)/\log n$ for $n\geq 2$ and $c(1)=0$. Define $\Pi(x)=\sum_{n\leq x}c(n)$ (so that Π corresponds to ψ in the same way that π corresponds to θ). Show that $c(p^k)=1/k$ for prime p, and hence that

$$\Pi(x)=\pi(x)+\frac{1}{2}\pi(x^{1/2})+\cdots+\frac{1}{m}\pi(x^{1/m}),$$

where m is the largest integer such that $2^m\leq x$. Apply Abel's summation formula to express $\Pi(x)$ in terms of $\psi(x)$, and $\psi(x)$ in terms of $\Pi(x)$.

6 By considering numbers of the form $n=p_1p_2\ldots p_m$, the product of the first m primes, show that $\tau(n)$ is not $O[(\log n)^k]$ for any k. (Apply Chebyshev's lower estimate to $m=\pi(p_m)$ and upper estimate to $\log n=\theta(p_m)$.)

7 Let p_n denote the nth prime. Show that there are constants a, A such that

$$an\log n\leq p_n\leq An\log n$$

for all $n\geq 2$. (You will need a statement like: if $x\leq y\log x$, then $x\leq 2y\log y$. To prove this, recall that $\log u<\frac{1}{2}u$.)

2.5 Estimates of some summation functions

The next two sections could be deferred by readers wanting a fast route to the prime number theorem, since they are not used in the proof. However, they serve to provide a context for it, showing how it can be seen as one of a family of theorems of a similar type. We have seen that the prime number theorem is equivalent to a certain estimation of $\psi(x)$, the summation function of von Mangoldt's Λ. It is natural to consider the summation functions of other standard arithmetic functions. Happily, most of them can be estimated a lot more easily than $\psi(x)$. In this section, we shall present the results for four of the best-known cases. As the reader will see, the methods vary considerably from case to case, but a common feature is the repeated use of integrals to estimate sums.

From now on, we shall occasionally use the O, o notation (see appendix E), but when it is easy to give definite values for the constants appearing in estimations, we shall usually state the result in a form that incorporates these values.

The divisor function

We saw in 1.4.5 that $\sum_{n \le x} \tau(n)$ lies between $x(\log x - 1)$ and $x(\log x + 1)$, which is already enough to show that, in a sense, the "average" value of $\tau(n)$ is $\log n$ (see exercise 6). We now present a more accurate estimate, first obtained by Dirichlet in 1841. The new ingredients are (i) Dirichlet's hyperbola identity (1.8.6), requiring summation only up to $x^{1/2}$, and (ii) the better estimation of $\sum_{j \le x} \frac{1}{j}$ derived from Euler's summation formula.

Proposition 2.5.1 *For all $x \ge 1$,*

$$\sum_{n \le x} \tau(n) = x \log x + (2\gamma - 1)x + q(x),$$

where $|q(x)| \le 4x^{1/2}$.

Proof Denote the sum by $S(x)$. Recall that $\tau = u * u$. Apply 1.8.6 with $a = b = u$ and $y = x^{1/2}$. Note that $A(x) = B(x) = [x]$. We obtain $S(x) = S_1 + S_2 - S_0$, where

$$S_1 = S_2 = \sum_{j \le x^{1/2}} \left[\frac{x}{j} \right]$$

and $S_0 = [x^{1/2}]^2$. (It is not hard to arrive at this formula without the language of convolutions, by equating $S(x)$ to the number of lattice points

(j,k) with $j,k \geq 1$ and $jk \leq x$; a "lattice point" means an ordered pair of integers.)

Since $x^{1/2} - 1 \leq [x^{1/2}] \leq x^{1/2}$, we have $S_0 = x - q_1(x)$, where $0 \leq q_1(x) \leq 2x^{1/2}$.

We now estimate S_1. Recall that $\{x\}$ denotes $x - [x]$, the "fractional part" of x. Clearly, $0 \leq \{x\} < 1$. So

$$\begin{aligned} S_1 &= \sum_{j \leq x^{1/2}} \left(\frac{x}{j} - \left\{ \frac{x}{j} \right\} \right) \\ &= x \sum_{j \leq x^{1/2}} \frac{1}{j} - q_2(x), \end{aligned}$$

where $0 \leq q_2(x) \leq x^{1/2}$. Further, by 1.4.11,

$$x \sum_{j \leq x^{1/2}} \frac{1}{j} = x(\tfrac{1}{2} \log x + \gamma) + q_3(x),$$

where $|q_3(x)| \leq x^{1/2}$. So, finally,

$$S(x) = 2S_1 - S_0 = x \log x + (2\gamma - 1)x + 2q_3(x) - 2q_2(x) + q_1(x).$$

Clearly, $|q_1(x) - 2q_2(x)| \leq 2x^{1/2}$. The statement follows. □

The expression for $S(x)$ used in the proof is also useful for the computation of it for particular values of x. The following table compares some actual values of the sum with the estimate $x \log x + (2\gamma - 1)x$.

x	100	1,000	10,000	100,000	1,000,000
actual	482	7,069	93,668	1,166,750	13,970,034
estimate	476	7,062	93,648	1,166,736	13,969,942

At least for these values, the difference $q(x)$ is visibly smaller than the estimate $4x^{1/2}$ given by 2.5.1. The problem of determining the true order of magnitude of $q(x)$ is called the "Dirichlet divisor problem", and it has been the subject of a great deal of study. Denote by θ_0 the infimum of numbers θ such that $q(x)$ is $O(x^\theta)$. Voronoi already showed in 1903 that $\theta_0 \leq 1/3$ (see, e.g., [Ten]). Currently the best estimate, due to Huxley (1993), is $\theta_0 \leq \frac{23}{73}$. At the same time, it is known that $\theta_0 \geq \frac{1}{4}$.

Representations as the sum of two squares

This result, which is due to Gauss, is an excellent example of the solution of a number-theoretic problem by geometric ideas. Observe first that

$$325 = 15^2 + 10^2 = 17^2 + 6^2 = 18^2 + 1^2.$$

We define $r(n)$ to be the number of ordered pairs (j, k) of integers such that $n = j^2 + k^2$. This distinguishes (k, j) from (j, k) if $j \neq k$. It also allows negative integers and 0, so in the case when j, k are distinct and non-zero, the combinations $(\pm j, \pm k)$ and $(\pm k, \pm j)$ count 8 towards $r(n)$. Hence, for example, $r(325) = 24$. (The reader may be inclined to divide by 8 in the result that follows). Of course, in many cases $r(n) = 0$.

Write $D(\rho)$ for the disc with centre $(0, 0)$ and radius ρ, in other words, the set of points (ξ, η) with $\xi^2 + \eta^2 \leq \rho^2$. If $n \leq x$ and $n = j^2 + k^2$, then (j, k) is in $D(x^{1/2})$. Hence $\sum_{n \leq x} r(n)$ is the number of lattice points (j, k) in this disc (we are including $n = 0$, with $r(0) = 1$). It seems almost obvious that the number of such points approximates to the area of the disc, πx. The next result makes this precise.

Proposition 2.5.2 *For all $x > 1$, we have*

$$\sum_{n \leq x} r(n) = \pi x + q(x),$$

where $|q(x)| \leq \pi(2x)^{1/2} + 2$.

Proof Let $S(j, k)$ be the square with side length 1 and centre at (j, k). Such squares clearly form a grid covering the plane. Let S be the union of all the squares $S(j, k)$ for lattice points (j, k) with $j^2 + k^2 \leq x$. By the remarks above, the area of S is $\sum_{n \leq x} r(n)$. We will estimate this area by identifying discs containing S and contained in S.

Let the distance from a lattice point (j, k) to the origin be ρ. Any point of $S(j, k)$ is within distance $1/\sqrt{2}$ of (j, k), so, by the triangle inequality for distances, the distance from any such point to the origin is between $\rho - 1/\sqrt{2}$ and $\rho + 1/\sqrt{2}$. (It is a good exercise to give an algebraic proof of this statement.) It follows at once that S is contained in $D(x^{1/2} + 1/\sqrt{2})$. On the other hand, choose a point (ξ, η) of $D(x^{1/2} - 1/\sqrt{2})$. Then (ξ, η) is in $S(j, k)$ for some (j, k), and, by the above, this (j, k) is in $D(x^{1/2})$. So this second disc is contained in S. Hence the area of S lies between the areas of these two discs, which are

$$\pi(x^{1/2} \pm 1/\sqrt{2})^2 = \pi\left(x \pm (2x)^{1/2} + \tfrac{1}{2}\right). \qquad \square$$

The determination of the true order of magnitude of the error term $q(x)$ in 2.5.2 is the "Gauss circle problem". Known estimations are very similar to those obtained for the Dirichlet divisor problem. Let θ_1 denote the infimum of θ such that $q(x) = O(x^\theta)$. Again we have $\theta_1 \geq \frac{1}{4}$, together with Huxley's estimate $\theta_1 \leq \frac{23}{73}$.

Euler's function ϕ

We now consider Euler's ϕ function, which was discussed in section 2.2. The method is rather more typical of those used for problems of this kind. We again use integral estimation, together with several of our earlier results, including the convolution identity $\phi = I * \mu$ and the series

$$\sum_{k=1}^{\infty} \frac{\mu(k)}{k^2} = \frac{1}{\zeta(2)} = \frac{6}{\pi^2}. \tag{2.1}$$

Proposition 2.5.3 *For Euler's function ϕ, we have*

$$\sum_{n \leq x} \phi(n) = \frac{3}{\pi^2} x^2 + q(x),$$

where $|q(x)| \leq x(\log x + 2)$.

Proof Write $\sum_{n \leq x} \phi(n) = S(x)$. By 2.2.16, we have $\phi = I * \mu$, where $I(n) = n$. So, by 1.8.4,

$$S(x) = \sum_{k \leq x} \mu(k) \sum_{j \leq x/k} j.$$

By 1.4.3,

$$\sum_{j \leq x/k} j = \int_0^{x/k} t \, dt + a_k = \frac{x^2}{2k^2} + a_k,$$

where $|a_k| \leq (x/k)$. So

$$S(x) = \tfrac{1}{2} x^2 \sum_{k \leq x} \frac{\mu(k)}{k^2} + q_1(x),$$

where, by 1.4.2,

$$|q_1(x)| \leq x \sum_{k \leq x} \frac{1}{k} \leq x(\log x + 1).$$

Now, by 1.4.4,

$$\left|\sum_{k>x}\frac{\mu(k)}{k^2}\right| \leq \sum_{k>x}\frac{1}{k^2} \leq \int_x^\infty \frac{1}{t^2}\,dt + \frac{1}{x^2} = \frac{1}{x} + \frac{1}{x^2} \leq \frac{2}{x}.$$

So, by (2.1),

$$\sum_{k\leq x}\frac{\mu(k)}{k^2} = \frac{6}{\pi^2} + p(x),$$

where $|p(x)| \leq 2/x$. Hence

$$S(x) = \frac{3}{\pi^2}x^2 + q_1(x) + q_2(x),$$

where $|q_2(x)| = \frac{1}{2}x^2|p(x)| \leq x$. The statement follows. □

This result has an interesting interpretation. There are N^2 lattice points (m,n) with $1 \leq m,n \leq N$. Of these, the number satisfying $(m,n) = 1$ is $2\sum_{n\leq N}\phi(n)$: the factor 2 arises because there are equally many such points with $m < n$ and with $m > n$. Our result says that, for large N, the *proportion* of lattice points having $(m,n) = 1$ approximates to $6/\pi^2$. Roughly speaking, this is the probability that a randomly chosen pair of numbers is coprime. Note that the condition $(m,n) = 1$ also has a simple geometric meaning: it equates to (m,n) being "visible" from the origin, in the sense that no other lattice point lies between them.

Again we illustrate the result with some actual values:

x	100	1,000	10,000
actual	3,044	304,192	30,397,486
estimate	3,040	303,964	30,396,355

Only slight improvements to the error estimate in 2.5.3 have been obtained, involving the power of $\log x$. Since the summation function has a jump of $n-1$ at prime n, the order of $q(x)$ is at least x. Computations show that $|q(n)| < n$ for all integers $n \leq 10^6$.

μ^2 *and the proportion of square-free numbers*

Recall that $\mu(n)^2$ $(= |\mu(n)|)$ is 1 when n is square-free and 0 otherwise. So its summation function is simply the number of square-free integers not greater than x. To estimate it, we shall again use the series for $1/\zeta(2)$.

Lemma 2.5.4 *For each n, we have* $\mu(n)^2 = \sum_{m^2|n}\mu(m)$.

Proof Write $S(n) = \sum_{m^2|n} \mu(m)$. If n is square-free, then 1 is the only m such that $m^2|n$, so $S(n) = \mu(1) = 1$.

Now suppose that n is not square-free. We have to show that $S(n) = 0$. From the prime factorization of n, it is clear that n can be expressed as h^2k, where $h > 1$ and k is square-free. Further, it is clear that if $m^2|n$, then $m|h$. So, by 2.2.5,

$$S(n) = \sum_{m|h} \mu(m) = 0. \qquad \square$$

Proposition 2.5.5 *Write* $M_2(x) = \sum_{n \le x} \mu(n)^2$. *Then, for all* $x \ge 1$,

$$M_2(x) = \frac{6}{\pi^2}x + q(x),$$

where $|q(x)| \le 2x^{1/2} + 1$. *Hence the proportion of square-free numbers in* $[1, x]$ *tends to* $6/\pi^2$ *as* $x \to \infty$.

Proof By the lemma,

$$M_2(x) = \sum_{n \le x} \sum_{m^2|n} \mu(m).$$

For each $m \le x^{1/2}$, the term $\mu(m)$ appears in this sum $[x/m^2]$ times, so

$$\begin{aligned} M_2(x) &= \sum_{m \le x^{1/2}} \mu(m) \left[\frac{x}{m^2}\right] \\ &= S_1(x) + S_2(x), \end{aligned}$$

where

$$S_1(x) = x \sum_{m \le x^{1/2}} \frac{\mu(m)}{m^2}, \qquad S_2(x) = - \sum_{m \le x^{1/2}} \mu(m) \left\{\frac{x}{m^2}\right\}.$$

Clearly, $|S_2(x)| \le x^{1/2}$. As in the proof of 2.5.3,

$$\sum_{m \le x^{1/2}} \frac{\mu(m)}{m^2} = \frac{6}{\pi^2} + q_1(x),$$

where

$$|q_1(x)| \le \sum_{m > x^{1/2}} \frac{1}{m^2} \le \frac{1}{x^{1/2}} + \frac{1}{x}.$$

Hence

$$S_1(x) + \frac{6}{\pi^2}x + q_2(x),$$

where $|q_2(x)| \le x^{1/2} + 1$. The statement follows. $\square$

The distribution of square-free numbers appears to be extremely uniform right from the start. The numbers of them occurring in the first 10 intervals of length 100 are as follows:

$$61 \quad 61 \quad 61 \quad 60 \quad 63 \quad 60 \quad 62 \quad 61 \quad 58 \quad 61.$$

Hence 608 of the numbers from 1 to 1000 are square-free. By comparison, $6/\pi^2 \approx 0.607927$. However, we shall see in section 5.2 that $q(x)$ is at least $O(x^{1/4})$.

The reader may wonder why we have given an estimate for the summation function of μ^2 but not of μ itself. The answer is that this is much harder! It will be partially solved, in the sense of showing that $M(x)/x \to 0$ as $x \to \infty$, at the same time as proving the prime number theorem. But the determination of the true rate of growth of $M(x)$ is one of the great unsolved problems in the subject. We shall discuss it further in chapter 5.

Exercises

1 Use the expression in 2.5.1 to show that $\sum_{n \le 100} \tau(n) = 482$. (You should not need a calculator!)

2 Show that if $n \equiv 3 \pmod 4$, then $r(n) = 0$ (consider squares of even and odd numbers mod 4). Show also that if $r(m) > 0$ and $r(n) > 0$, then $r(mn) > 0$.

3 Show that

$$\sum_{n \le x} [\tau(n) - \log n - 2\gamma] = O(x^{1/2}).$$

4 Let $S_\tau(x) = \sum_{n \le x} \tau(n)$. Use the expression in 1.2.2 to show that

$$S_\tau(x) = x \log x + \gamma x - \sum_{n \le x} \left\{ \frac{x}{n} \right\} + O(1)$$

for all $x > 1$, and deduce that

$$\sum_{n \le x} \left\{ \frac{x}{n} \right\} = (1 - \gamma)x + O(x^{1/2}).$$

5 Use Abel summation to show that, for certain constants c_0, c_1,

$$\sum_{n \le x} \frac{\tau(n)}{n} = \tfrac{1}{2}(\log x)^2 + 2\gamma \log x + c_0 + O(x^{-1/2}),$$

$$\sum_{n \le x} \frac{\mu(n)^2}{n} = \frac{6}{\pi^2} \log x + c_1 + O(x^{-1/2}).$$

6 Prove that

$$\sum_{2\leq n\leq x} \frac{\tau(n)}{\log n} = x + O\left(\frac{x}{\log x}\right)$$

and

$$\sum_{n\leq x} \frac{\phi(n)}{n} = \frac{6}{\pi^2}x + O(\log^2 x)$$

(so that the average values of $\tau(n)/\log n$ and $\phi(n)/n$ tend to 1 and $6/\pi^2$, respectively).

7 Define $\sigma_1(n) = \sum_{j|n} j$ and $S_1(x) = \sum_{n\leq x} \sigma_1(x)$. Show that

$$S_1(x) = \sum_{k\leq x} \sum_{j\leq x/k} j$$

and deduce that

$$S_1(x) = \frac{\pi^2}{12}x^2 + q(x),$$

where $|q(x)| \leq x(\log x + 2)$.

2.6 Mertens's estimates

To prove the prime number theorem, as we have seen, we need to estimate the summation function of $\Lambda(n)$. Here we show that it is a good deal easier to do the same for $\Lambda(n)/n$, as one might expect since the terms are smaller. The result then translates into estimates of $\sum_{p\in P[x]}(\log p/p)$ and $\sum_{p\in P[x]}(1/p)$, which were first obtained by F. Mertens in 1874. We already have some estimates for the second case (1.6.5 and 2.1.6), but our new estimates will be much more accurate. The section finishes with some applications of these theorems, such as the estimation of the summation function of $\omega(n)$, the number of prime factors of n. The methods will be essentially number-theoretic: the main ingredients are Chebyshev's upper estimate, the convolution identity $\Lambda * u = \ell$ and (of course) integral estimation.

These results are not needed for either variant of our main proof of the prime number theorem in chapter 3, so the reader could omit this section. However, the theorems are very much part of the story, both historically and because they deal with variations of the same problem. They are more directly related to prime numbers than the results of section 2.5. Also, the first theorem forms a basic step in the "elementary" proof of the prime number theorem in chapter 6.

Partial sums of $\sum \Lambda(n)/n$ *and* $\sum(\log p)/p$

The proof of our first result is along rather similar lines to the proof that the partial sums of $\sum \mu(n)/n$ are bounded (2.2.9).

Theorem 2.6.1 *For all* $x > 1$, *we have*

$$\sum_{n \le x} \frac{\Lambda(n)}{n} = \log x + r(x),$$

where $|r(x)| \le 2$.

Proof Write $S(x) = \sum_{n \le x} \log n$. Since $\Lambda * u = \ell$, we have, by 1.8.5,

$$\begin{aligned} S(x) &= \sum_{n \le x} (\Lambda * u)(n) \\ &= \sum_{n \le x} \Lambda(n) \left[\frac{x}{n}\right] \\ &= x \sum_{n \le x} \frac{\Lambda(n)}{n} - a(x), \end{aligned} \tag{2.2}$$

where

$$a(x) = \sum_{n \le x} \Lambda(n) \left\{\frac{x}{n}\right\}.$$

By Chebyshev's upper estimate $\psi(x) \le 2x$ (see 2.4.6),

$$0 \le a(x) \le \sum_{n \le x} \Lambda(n) = \psi(x) \le 2x.$$

We now use 1.4.3 to estimate $S(x)$. Since

$$\int_1^x \log t \, dt = x \log x - x + 1,$$

we have

$$S(x) = x \log x - x + b(x),$$

where $|b(x)| \le \log x + 1$. It is elementary that $\log x + 1 \le x$ for $x > 1$. By (2.2), we now have

$$\begin{aligned} x \sum_{n \le x} \frac{\Lambda(n)}{n} &= S(x) + a(x) \\ &= x \log x - x + a(x) + b(x). \end{aligned}$$

Clearly, $|a(x) - x + b(x)| \le 2x$. The statement follows. □

This theorem lends some support to our basic conjecture that $\Lambda(n)$, when averaged out, is like 1. Since $\log x = \sum_{n \leq x} \frac{1}{n} + O(1)$, an equivalent statement of the result is that the partial sums of the series $\sum_{n=1}^{\infty} [\Lambda(n) - 1]/n$ are bounded. As we shall see later, this series is actually convergent, and this statement is essentially equivalent to the prime number theorem.

Corollary 2.6.2 *We have*

$$\int_1^x \frac{\psi(t)}{t^2}\, dt = \log x + O(1) \qquad \textit{for } x > 1.$$

Proof By 1.3.6, we have

$$\sum_{n \leq x} \frac{\Lambda(n)}{n} = \frac{\psi(x)}{x} + \int_1^x \frac{\psi(t)}{t^2}\, dt.$$

Since $\psi(x)/x$ is bounded, the statement follows. □

It is easy to deduce that if $\psi(t)/t$ tends to any limit as $t \to \infty$, then the limit is 1. For if $h(t) \to c$ as $t \to \infty$, then it is not hard to show that

$$\frac{1}{\log x} \int_1^x \frac{h(t)}{t}\, dt \to c \quad \text{as } x \to \infty.$$

But 2.6.2 shows that when $h(t)$ is $\psi(t)/t$, this statement holds with $c = 1$.

The next result, Mertens's "first theorem", is a variant of 2.6.1 with the higher powers of primes removed, leaving us with $\sum_{p \in P[x]} (\log p/p)$. The point is that the powers of primes are so sparse that this has no effect on the estimate.

Proposition 2.6.3 *We have*

$$\sum_{p \in P[x]} \frac{\log p}{p} = \log x + O(1) \qquad \textit{for } x > 1.$$

Proof Recall that $\Lambda(n) = \log p$ when $n = p^m$. It follows that

$$\begin{aligned} 0 \leq \sum_{n \leq x} \frac{\Lambda(n)}{n} - \sum_{p \in P[x]} \frac{\log p}{p} &\leq \sum_{p \in P[x]} \log p \left(\frac{1}{p^2} + \frac{1}{p^3} + \cdots \right) \\ &= \sum_{p \in P[x]} \frac{\log p}{p(p-1)} \\ &\leq \sum_{n=2}^{\infty} \frac{\log n}{n(n-1)}. \end{aligned}$$

This series is convergent (with sum less than 2). The statement follows from 2.6.1 (and the constant in the $O(1)$ is certainly no more than 4). □

Partial sums of $\sum 1/p$ and some applications

We now apply Abel summation to deduce an estimate for $\sum_{p\in P[x]}(1/p)$: this is Mertens's "second theorem". The terms of the series are smaller by the factor $\log p$, which has the effect that we can show that the difference between the sum and $\log\log x$ tends to a limit (just as $\sum_{n\leq x}(1/n) - \log x$ tends to Euler's constant).

Theorem 2.6.4 *There is a constant C_1 such that*

$$\sum_{p\in P[x]} \frac{1}{p} = \log\log x + C_1 + O\left(\frac{1}{\log x}\right) \quad \textit{for all } x \geq 2.$$

Proof Define

$$a(n) = \begin{cases} (\log n)/n & \text{if } n \text{ is prime,} \\ 0 & \text{otherwise.} \end{cases}$$

Then

$$\sum_{p\in P[x]} \frac{\log p}{p} = A(x),$$

where (as usual) we write $A(x) = \sum_{n\leq x} a(n)$. By 2.6.3, $A(x) = \log x + r(x)$, where $|r(x)| \leq c_0$ (say) for all $x > 1$. Clearly,

$$\sum_{p\in P[x]} \frac{1}{p} = \sum_{2\leq n\leq x} \frac{a(n)}{\log n}.$$

Since $a(1) = 0$, we can apply version 1.3.7 of Abel's summation formula to obtain

$$\begin{aligned} \sum_{p\in P[x]} \frac{1}{p} &= \frac{A(x)}{\log x} + \int_2^x \frac{A(t)}{t(\log t)^2}\, dt \\ &= 1 + \frac{r(x)}{\log x} + \int_2^x \frac{1}{t\log t}\, dt + I(x), \end{aligned}$$

where

$$I(x) = \int_2^x \frac{r(t)}{t(\log t)^2}\, dt.$$

Now

$$\int_2^x \frac{1}{t\log t}\,dt = \log\log x - \log\log 2,$$

and

$$\int_2^\infty \frac{1}{t(\log t)^2}\,dt$$

is convergent, since

$$\frac{1}{t(\log t)^2} = -\frac{d}{dt}\,\frac{1}{\log t}.$$

It follows that $I(x)$ tends to a limit (say I) as $x \to \infty$, and $I(x) = I - s(x)$, where

$$|s(x)| \le c_0 \int_x^\infty \frac{1}{t(\log t)^2}\,dt = \frac{c_0}{\log x}.$$

The statement follows, with $C_1 = 1 - \log\log 2 + I$. □

This proof does not give much idea of the value of the constant C_1, but a more detailed analysis leads to an explicit formula for it, giving the value $C_1 \approx 0.26150$ (see the exercises). As an example of actual values, we find by calculation:

$$\sum_{p\in P[100]} \frac{1}{p} \approx 1.8029, \qquad \log\log 100 \approx 1.5272.$$

Next, we extend this result to the case where all powers of primes are included, instead of just primes. This might seem a rather unnatural sum to consider, but it is exactly what is needed for the application that follows. Let $P^*[x]$ be the set of all numbers not greater than x that are of the form p^k for some prime p and some integer $k \ge 1$. As before, we write $\pi^*(x)$ for the number of members of $P^*[x]$. Recall from 2.4.7 that $\pi^*(x) - \pi(x) = O(x^{1/2}/\log x)$, so that $\pi^*(x)$, like $\pi(x)$, is $O(x/\log x)$.

Proposition 2.6.5 *For all $x \ge 2$,*

$$\sum_{q\in P^*[x]} \frac{1}{q} = \log\log x + C_2 + O\left(\frac{1}{\log x}\right),$$

in which $C_2 = C_1 + S$, where C_1 is the constant in 2.6.4 and

$$S = \sum_{p\in P} \frac{1}{p(p-1)}.$$

(In fact, $C_2 \approx 1.03465$.)

Proof Write

$$\Delta = \sum_{q \in P^*[x]} \frac{1}{q} - \sum_{p \in P[x]} \frac{1}{p}.$$

For each prime p such that $p^2 \leq x$, let r_p be the largest r such that $p^r \leq x$. Then we have

$$\Delta = \sum_{p \in P[x^{1/2}]} \sum_{r=2}^{r_p} \frac{1}{p^r}.$$

By the geometric series,

$$\sum_{r=2}^{\infty} \frac{1}{p^r} = \frac{1}{p(p-1)}.$$

Hence $\Delta \leq S$. We will show that $S - \Delta \leq (4/x^{1/2})$; the result then follows from 2.6.4. Let

$$S_0 = \sum_{p \in P[x^{1/2}]} \frac{1}{p(p-1)}.$$

Then

$$S - S_0 < \sum_{n > x^{1/2}} \frac{1}{n(n-1)} = \frac{1}{[x^{1/2}]} \leq \frac{2}{x^{1/2}}. \tag{2.3}$$

Also, since $p^{r_p+1} > x$, we have

$$\sum_{r > r_p} \frac{1}{p^r} < \frac{1}{x}\left(1 + \frac{1}{p} + \frac{1}{p^2} + \cdots\right) = \frac{1}{x\,(1 - 1/p)} \leq \frac{2}{x}.$$

So

$$S_0 - \Delta = \sum_{p \in P[x^{1/2}]} \sum_{r > r_p} \frac{1}{p^r} < \frac{2}{x}\pi(x^{1/2}) < \frac{2}{x^{1/2}}. \tag{2.4}$$

By (2.3) and (2.4), we have $S - \Delta \leq (4/x^{1/2})$, as stated. □

We now apply the last two results to estimate the summation functions of ω and Ω.

Proposition 2.6.6 *For all $x \geq 2$,*

$$\sum_{n \leq x} \omega(n) = x \log\log x + C_1 x + O\left(\frac{x}{\log x}\right),$$

$$\sum_{n \leq x} \Omega(n) = x \log\log x + C_2 x + O\left(\frac{x}{\log x}\right),$$

where C_1 and C_2 are the constants in 2.6.4 and 2.6.5.

Proof Recall from 1.2.2 that

$$\sum_{n \le x} \omega(n) = \sum_{p \in P[x]} \left[\frac{x}{p}\right] = \sum_{p \in P[x]} \left(\frac{x}{p} - \left\{\frac{x}{p}\right\}\right).$$

By 2.6.4,

$$\sum_{p \in P[x]} \frac{x}{p} = x \sum_{p \in P[x]} \frac{1}{p} = x \log\log x + C_1 x + O\left(\frac{x}{\log x}\right).$$

Also,

$$0 \le \sum_{p \in P[x]} \left\{\frac{x}{p}\right\} \le \pi(x),$$

and by Chebyshev's estimate, $\pi(x) = O(x/\log x)$ for all $x \ge 2$.

Now consider the second sum. Note that $\Omega(n)$ is the number of pairs (p^k, n) with p prime, $k \ge 1$ and $p^k | n$, since if $p_j^{k_j}$ appears in the prime factorization of n, then there are k_j such pairs with $p = p_j$. For fixed $q = p^k$, there are $[x/q]$ such pairs with $n \le x$. Hence

$$\sum_{n \le x} \Omega(n) = \sum_{q \in P^*[x]} \left[\frac{x}{q}\right].$$

The statement now follows in the same way as for the first sum, using 2.6.5 and the fact, noted above, that $\pi^*(x)$ is also $O(x/\log x)$. □

From the way in which the terms $\{x/p\}$ were discarded, it is fairly clear that the error term will be of the order of magnitude stated. Some values are

x	100	1,000	10,000	100,000	1,000,000
actual (ω)	171	2,126	24,300	266,400	2,853,708
estimate	179	2,194	24,818	270,497	2,887,289
actual (Ω)	239	2,877	31,985	343,614	3,626,619
estimate	256	2,967	32,550	347,812	3,660,445

A second application concerns the enumeration of numbers with a "large" prime factor. Denote by $P(x, y]$ the set of primes p such that $x < p \le y$. As an immediate consequence of 2.6.4 (in which C_1 disappears by cancellation), we have:

Proposition 2.6.7 *Let $0 < \alpha < 1$. Then*

$$\sum_{p\in P(x^\alpha,x]} \frac{1}{p} = -\log\alpha + O\left(\frac{1}{\log x}\right) \qquad \textit{for } x \geq 2.$$

(Note that $\log\alpha < 0$.)

Proof This follows from 2.6.4 and the simple identity

$$\log\log x - \log\log x^\alpha = \log\log x - \log(\alpha\log x) = -\log\alpha. \qquad \square$$

Proposition 2.6.8 *Let $\frac{1}{2} \leq \alpha < 1$. Let $S(x,\alpha)$ be the set of integers not greater than x that have a prime factor larger than x^α, and let $f(x,\alpha)$ be the number of members of $S(x,\alpha)$. Then*

$$f(x,\alpha) = -x\log\alpha + O\left(\frac{x}{\log x}\right) \qquad \textit{for } x > 2.$$

Proof Consider the set of pairs (p,n) with $p|n$, $n \leq x$ and $p > x^\alpha$. Each member n of $S(x,\alpha)$ appears in exactly one such pair, since the condition $\alpha \geq \frac{1}{2}$ ensures that n does not have two prime factors bigger than x^α. Hence the number of such pairs is $f(x,\alpha)$, so, by 2.6.7,

$$\begin{aligned} f(x,\alpha) &= \sum_{p\in P(x^\alpha,x]} \left[\frac{x}{p}\right] \\ &= \sum_{p\in P(x^\alpha,x]} \left(\frac{x}{p} - \left\{\frac{x}{p}\right\}\right) \\ &= x\sum_{p\in P(x^\alpha,x]} \frac{1}{p} + O(\pi(x)) \\ &= -x\log\alpha + O\left(\frac{x}{\log x}\right). \end{aligned} \qquad \square$$

Exercises

1 Let $R(x) = \prod_{p\in P[x]}(1-1/p)^{-1}$. Let C_1 be the constant in 2.6.4 and

$$C_0 = \sum_{p\in P}\left[-\log\left(1-\frac{1}{p}\right) - \frac{1}{p}\right].$$

Use the method of 2.1.6 to estimate the tail of the series for C_0, and apply the results of this section to show that

$$R(x) = e^{C_0+C_1}\log x + O(1).$$

(It will help to recall that $e^y = 1 + h(y)$, where $|h(y)| \le 2|y|$ when $|y| \le \frac{1}{2}$.) *Note: One finds that $C_0 \approx 0.3157$; it can be shown that $C_0 + C_1 = \gamma$ (see [Ten], section 1.6).*

2 Let the integer n (≥ 3) have m prime divisors, and let p_m be the mth prime. Use the previous exercise in the weaker form $R(x) \le K \log x$ to show that $\phi(n) \ge n/(K \log p_m)$. Now use Chebyshev's lower estimate to show that for certain constants A, C,

$$\phi(n) \ge \frac{Cn}{\log\log n + A} \qquad (n \ge 3).$$

3 (*Upper bounds for $\omega(n)$*) Let $n_k = p_1 p_2 \ldots p_k$, the product of the first k primes, and write

$$g(n) = \frac{\log n}{\log\log n}.$$

Show that if $\omega(n) = k$, then $n \ge n_k$. Prove that, for any $\varepsilon > 0$, we have $\omega(n_k) > (1 - \varepsilon) g(n_k)$ for all sufficiently large k. [Note that $\log n_k = \theta(p_k)$ and also $\log n_k \le k \log p_k$.]

Now use Chebyshev's estimates for $\pi(p_k)$ $(= k)$ and $\theta(p_k)$, together with the fact that $x/(\log x)$ increases for $x \ge e$, to prove that there is a constant C such that $\omega(n) \le C g(n)$ for all $n \ge 3$.

3

The basic theorems

Recall that our overall strategy is based on the series $-\zeta'(s)/\zeta(s) = \sum_{n=1}^{\infty} \Lambda(n)/n^s$: the hope is to deduce the required estimation of $\psi(x)$ from the properties of the function $\zeta'(s)/\zeta(s)$. This procedure can be carried out for a general Dirichlet series: given that $\sum_{n=1}^{\infty} a(n)/n^s$ converges (on a suitable domain) to a function $f(s)$, we shall show that if $f(s)$ has certain properties, then $A(x)/x$ tends to a limit. This is our "fundamental theorem". In this way, we finally obtain the prime number theorem as one among a family of theorems. Though it is certainly the most important one, other theorems in the family are of considerable interest; we shall meet some later in this chapter and more in chapter 4.

The fundamental theorem will actually be obtained in three different versions, describing respectively an integral, a limit (as just mentioned) and a series. The integral version will be proved first, and the other two will be derived from it. An example of the type of series obtained is $\sum_{n=1}^{\infty} \mu(n)/n = 0$.

Though we have taken care to consider the zeta function, and other Dirichlet series, as functions of a *complex* variable s, we have not really made any use of this fact so far. Complex numbers were not used in proving any of the results in chapters 1 and 2. However, the fundamental theorem depends in an essential way on the properties of $f(s)$ as a *complex* function. Very roughly, the method is based on the idea of "inverting" the Dirichlet series to express $A(x)$ in terms of $f(s)$: this is done by an integral on a vertical line in the complex plane. As with power series, no such expression is available within real analysis.

The fundamental theorem requires $f(s)$ to exist, and to be holomorphic, on a region including the line $\operatorname{Re} s = 1$ (except at the point 1). However, so far we have only defined $\zeta(s)$ for $\operatorname{Re} s > 1$. So our first task is to find a way of extending the definition of $\zeta(s)$ to a larger domain (for example, $\operatorname{Re} s > 0$). The extended function must be holomorphic, so that the theorems of complex analysis apply. It must also be *non-zero* on a region including Re $s = 1$, since the fundamental theorem is to be applied to $\zeta'(s)/\zeta(s)$, not $\zeta(s)$ itself. This extension is an interesting piece of work in its own right; it is discussed in section 3.1.

For the fundamental theorem itself, we then describe two alternative methods in sections 3.2 and 3.3. The first, a variant of the "traditional" method, is along the lines just mentioned. The second method was devised by D. J. Newman in 1980. It is slightly shorter, but less transparent, depending on an ingeniously chosen function. It still uses complex integrals. The two versions of the theorem are not identical, because one of the hypotheses is different in the two cases. The first method needs an estimation of $|f(1+it)|$ in terms of t, while Newman's method needs the condition $|A(x)| \leq Cx$ (in the case of the prime number theorem, this means Chebyshev's upper estimate). The first method is slightly preferable for the purpose of proving a more exact version of the prime number theorem with an error estimate (the subject of chapter 5). However, for the purposes of the present chapter (and indeed chapter 4), the reader may opt for either section 3.2 or section 3.3; later results are presented in a way that caters for either. Better still, read both and make comparisons!

Given the effort invested in proving the prime number theorem, one would hope to see some interesting applications. Some are described in section 3.5. For example, we show how to derive, for each k, an estimation of the number of integers less than n having k prime factors.

3.1 Extension of the definition of the zeta function

Extension to Re $s > 0$

The series $\sum_{n=1}^{\infty}(1/n^s)$ only converges when Re $s > 1$. Our next objective is to find an expression that coincides with it for such s while making sense, and in fact defining a differentiable (alias holomorphic) function, for a wider range of values. The uniqueness theorem for complex functions tells us that if this can be done at all, then it can only be done in one way. Hence our extension, if we can find it, must be the unique "right" one, and we take it to be the definition of $\zeta(s)$ on the wider range. At the same time, the new expression will give useful information about $\zeta(s)$ even in the original range Re $s > 1$.

Such an expression has already been provided by Euler's summation formula! Take $f(x) = 1/x^s$. Then we have $\int_1^{\infty} f(x)\,dx = 1/(s-1)$ for

Re $s > 1$, and 1.4.9 says

$$\zeta(s) = \frac{1}{s-1} + 1 - s\int_1^\infty \frac{x-[x]}{x^{s+1}}\,dx. \tag{3.1}$$

Now $0 \le x - [x] < 1$ for all x, so, by 1.7.6, the Dirichlet integral

$$\int_1^\infty \frac{x-[x]}{x^{s+1}}\,dx$$

converges, say to $I(s)$, for all $s = \sigma + it$ with $\sigma > 0$, and $|I(s)| \le 1/\sigma$. Also, by 1.7.11, $I(s)$ is holomorphic for such s, with

$$I'(s) = -\int_1^\infty \frac{(x-[x])\log x}{x^{s+1}}\,dx.$$

So we *define* $\zeta(s)$ by formula (3.1) for all $s \ne 1$ with Re $s > 0$, and we can state:

Theorem 3.1.1 *The function $\zeta(s)$, defined by (3.1), is defined and holomorphic for all $s \ne 1$ with* Re $s > 0$, *and satisfies*

$$\zeta(s) = \frac{1}{s-1} + 1 + r_1(s),$$

where $|r_1(s)| \le |s|/\sigma$. The derivative is given by

$$\zeta'(s) = -\frac{1}{(s-1)^2} - \int_1^\infty \frac{x-[x]}{x^{s+1}}\,dx + s\int_1^\infty \frac{(x-[x])\log x}{x^{s+1}}\,dx.$$

Furthermore, $\zeta(s) - \dfrac{1}{s-1}$ is holomorphic at $s = 1$, so $\zeta(s)$ has a simple pole at 1.

Proof We need only remark that the expression for $\zeta'(s)$ is found by the usual rules for differentiation, given the above expression for $I'(s)$. □

A variant of the defining formula is obtained if we replace $x - [x]$ by $x - [x] - \frac{1}{2}$, which varies between $-\frac{1}{2}$ and $\frac{1}{2}$. Since

$$s\int_1^\infty \frac{\frac{1}{2}}{x^{s+1}}\,dx = \frac{1}{2},$$

we have:

Proposition 3.1.2 *For $s \ne 1$ with* Re $s > 0$, *we have*

$$\zeta(s) = \frac{1}{s-1} + \frac{1}{2} + r_1^*(s), \tag{3.2}$$

where

$$r_1^*(s) = -s \int_1^\infty \frac{x - [x] - \frac{1}{2}}{x^{s+1}}\, dx, \tag{3.3}$$

and $|r_1^*(s)| \leq |s|/2\sigma$. □

Of course, this is related to the second form of Euler's summation formula. As we will show later, the integral in (3.3) actually converges whenever $\operatorname{Re} s > -1$, thereby extending the definition of $\zeta(s)$ to this region.

Clearly, $\zeta(\bar{s}) = \overline{\zeta(s)}$, and $\zeta(\sigma)$ is real for real σ.

Beyond this, formula (3.1) is not very useful in revealing further properties of $\zeta(s)$, and no use at all for computing its value. For these purposes, we can do better by relating the extended function to partial sums of the original series $\sum 1/n^s$, as follows. Recall Euler's summation formula for finite sums:

$$\sum_{n=2}^{N} f(n) = \int_1^N f(x)\, dx + \int_1^N (x - [x]) f'(x)\, dx.$$

Applied to $f(x) = 1/x^s$ (where $s \neq 1$), with the term $f(1) = 1$ inserted, this says

$$\sum_{n=1}^{N} \frac{1}{n^s} = 1 + \frac{1}{s-1} - \frac{N^{1-s}}{s-1} - s \int_1^N \frac{x - [x]}{x^{s+1}}\, dx. \tag{3.4}$$

By subtracting (3.4) from (3.1), we obtain at once:

Theorem 3.1.3 *For all* $s \neq 1$ *with* $\operatorname{Re} s > 0$ *and all integers* $N \geq 1$,

$$\zeta(s) = \sum_{n=1}^{N} \frac{1}{n^s} + \frac{N^{1-s}}{s-1} + r_N(s), \tag{3.5}$$

where

$$r_N(s) = -s \int_N^\infty \frac{x - [x]}{x^{s+1}}\, dx, \tag{3.6}$$

and we have

$$|r_N(s)| \leq \frac{|s|}{\sigma N^\sigma}.$$

Hence

$$\zeta(s) = \lim_{N\to\infty} \left(\sum_{n=1}^{N} \frac{1}{n^s} + \frac{N^{1-s}}{s-1} \right).$$

Proof We only need to observe that

$$\left| \int_N^\infty \frac{x-[x]}{x^{s+1}}\,dx \right| \le \int_N^\infty \frac{1}{x^{\sigma+1}}\,dx = \frac{1}{\sigma N^\sigma}. \qquad \square$$

Remark 1 This result shows how the partial sums of $\sum(1/n^s)$ behave when Re $s = 1$. Indeed,

$$\sum_{n=1}^{N} \frac{1}{n^{1+it}} = \zeta(1+it) - \frac{1}{it}N^{-it} + r_N,$$

where $r_N \to 0$ as $N \to \infty$. Since $N^{-it} = e^{-it\log N}$, this shows that for large N the partial sums end up almost rotating around the circle with centre $\zeta(1+it)$ and radius $1/t$.

Remark 2 Theorem 3.1.3 expresses $\zeta(s)$ as the limit (uniform on suitable sets) of a sequence of functions that are clearly holomorphic except at $s = 1$. This gives a proof that $\zeta(s)$ is holomorphic that avoids 1.7.10 and the underlying theorem on differentiation under the integral sign. However, this approach does not lead, without further work, to the expression for $\zeta'(s)$ given in 3.1.1, and it fails to apply at $s = 1$.

Power series expressions with centre at 1

Proposition 3.1.4 *We have*

$$\zeta(s) - \frac{1}{s-1} \to \gamma \quad \text{as } s \to 1,$$

where γ is Euler's constant. On some disc with centre at $s = 1$, $\zeta(s)$ has a power series expansion of the form

$$\zeta(s) = \frac{1}{s-1} + \gamma + \sum_{n=1}^{\infty} c_n (s-1)^n.$$

Proof For $s \neq 1$ with Re $s > 0$, we have $\zeta(s) = 1/(s-1) + 1 - sI(s)$, where

$$I(s) = \int_1^\infty \frac{x-[x]}{x^{s+1}}\,dx.$$

We have seen that $I(s)$ is defined and holomorphic (hence continuous) at 1. Hence

$$\lim_{s\to 1}\left(\zeta(s) - \frac{1}{s-1}\right) = 1 - I(1) = 1 - \int_1^\infty \frac{x-[x]}{x^2}\,dx.$$

By 1.4.10, this equals γ. By the general theory of complex functions, it follows that $\zeta(s)$ has a power series of the form stated, converging at least when $|s-1|<1$. □

Clearly, we have also

$$\zeta'(s) = -\frac{1}{(s-1)^2} + c_1 + 2c_2(s-1) + \cdots.$$

Proposition 3.1.5 *On some disc with centre at $s = 1$, we have power series expressions of the form*

$$\frac{1}{\zeta(s)} = (s-1) - \gamma(s-1)^2 + \cdots,$$

$$\frac{\zeta'(s)}{\zeta(s)} = -\frac{1}{s-1} + \gamma + a_1(s-1) + \cdots.$$

Proof First, note that $\zeta(s)$ is of the form $g(s)/(s-1)$, where $g(s)$ is differentiable and non-zero at 1. It follows that

$$\frac{1}{\zeta(s)} = (s-1)\frac{1}{g(s)} \quad \text{and} \quad \frac{\zeta'(s)}{\zeta(s)} = \frac{g'(s)}{g(s)} - \frac{1}{s-1},$$

in which $1/g(s)$ is well-defined and differentiable at 1. By the theory of complex functions, this implies that $\zeta'(s)/\zeta(s)$ has a power series (on some disc with centre 1) of the form

$$\frac{\zeta'(s)}{\zeta(s)} = -\frac{1}{s-1} + a_0 + a_1(s-1) + \cdots.$$

The product of this series with the series for $\zeta(s)$ must equal the series for $\zeta'(s)$. In forming the product, powers of $(s-1)$ are collected in the usual way (as allowed by the theorem on multiplication of series). In particular, the $(s-1)^{-1}$ term gives $a_0 - \gamma = 0$, so the series is as stated. The proof for $1/\zeta(s)$ is similar (and simpler). □

Estimates in terms of t

Recall that $|\zeta(\sigma+it)| \le \zeta(\sigma)$ for $\sigma > 1$. (Incidentally, this is not true when $\sigma < 1$!) As well as extending the definition of $\zeta(s)$, the methods of this section enable us to give a bound in terms of t instead of σ (we assume $t > 0$, since $|\zeta(\sigma - it)| = |\zeta(\sigma+it)|$). By 3.1.2, we have the following elementary estimate when $|s-1| > 1$:

$$|\zeta(s)| \le \frac{3}{2} + \frac{|s|}{2\sigma} \le \frac{3}{2} + \frac{\sigma+t}{2\sigma} \le 2 + \frac{t}{2\sigma},$$

suggesting growth of the order of t itself. However, a much stronger estimate applies. Here we state it for the case $\sigma \geq 1$, but the method can be adapted for the case $\sigma < 1$.

Proposition 3.1.6 *When $\sigma \geq 1$ and $t \geq 2$, we have*

$$|\zeta(\sigma + it)| \leq \log t + 4.$$

Proof Recall from 3.1.3 that, for any positive integer N,

$$\zeta(s) = \sum_{n=1}^{N} \frac{1}{n^s} + \frac{N^{1-s}}{s-1} + r_N(s),$$

where

$$|r_N(s)| \leq \frac{|s|}{\sigma N^\sigma} \leq \left(1 + \frac{t}{\sigma}\right) \frac{1}{N^\sigma}.$$

Let $\sigma \geq 1$ and $t \geq 2$, and take $N = [t]$, so that $N \leq t < N + 1$. Then

$$\left| \sum_{n=1}^{N} \frac{1}{n^s} \right| \leq \sum_{n=1}^{N} \frac{1}{n} \leq \log N + 1 \leq \log t + 1,$$

$$\left| \frac{N^{1-s}}{s-1} \right| \leq \frac{1}{t} \qquad (\text{since } |s-1| \geq t),$$

$$|r_N(s)| \leq \frac{1+t}{N} \leq \frac{N+2}{N} \leq 2,$$

since $N \geq 2$. These estimates combine to give $|\zeta(s)| \leq \log t + 3\frac{1}{2}$. □

Note It can be shown that $|\zeta(1 + it)|$ is unbounded for $t \geq 1$, so the $\log t$ cannot simply be replaced by a constant.

We now give a corresponding estimate for $|\zeta'(s)|$ (this is needed for our first proof of the fundamental theorems, but not for Newman's method).

Lemma 3.1.7 *For $N \geq 2$,*

$$\sum_{n=1}^{N} \frac{\log n}{n} \leq \tfrac{1}{2}(\log N)^2 + \tfrac{1}{8}.$$

Proof Write $f(x) = (\log x)/x$. Then $f'(x) = (1 - \log x)/x^2 < 0$ for $x > e$, so $f(x)$ is decreasing for $x > e$, and by integral comparison as in 1.4.1,

$$\sum_{n=4}^{N} \frac{\log n}{n} \leq \int_3^N \frac{\log x}{x}\, dx = \tfrac{1}{2}(\log N)^2 - \tfrac{1}{2}(\log 3)^2.$$

The statement follows, since $\log 1 = 0$ and by actual computation $\frac{1}{2}\log 2 + \frac{1}{3}\log 3 - \frac{1}{2}(\log 3)^2 \approx 0.109 < \frac{1}{8}$ (also, the statement holds for $N = 2$). □

Proposition 3.1.8 *For $\sigma \geq 1$, $t \geq 2$, we have*

$$|\zeta'(\sigma + it)| \leq \tfrac{1}{2}(\log t + 3)^2.$$

Proof By differentiation of (3.5) and (3.6), we obtain

$$\zeta'(s) = -\sum_{n=1}^{N} \frac{\log n}{n^s} - \frac{N^{1-s}\log N}{s-1} - \frac{N^{1-s}}{(s-1)^2} - I_1(s) + sI_2(s),$$

where

$$I_1(s) = \int_N^\infty \frac{x - [x]}{x^{s+1}}\,dx, \qquad I_2(s) = \int_N^\infty \frac{(x - [x])\log x}{x^{s+1}}\,dx.$$

Though this looks complicated, it is easy to estimate the terms in the same way as before. Let $\sigma \geq 1$, $t \geq 2$ and $N = [t]$. First, by the lemma,

$$\left|\sum_{n=1}^{N} \frac{\log n}{n^s}\right| \leq \tfrac{1}{2}(\log N)^2 + \tfrac{1}{8} \leq \tfrac{1}{2}(\log t)^2 + \tfrac{1}{8}.$$

As shown in lemma 3.1.7, $\log x/x$ has its greatest value at e, where it equals $1/e < 1/2$. Hence

$$\left|\frac{N^{1-s}\log N)}{s-1}\right| \leq \frac{\log N}{t} \leq \frac{\log t}{t} < \frac{1}{2},$$

$$\left|\frac{N^{1-s}}{(s-1)^2}\right| \leq \frac{1}{t^2} \leq \frac{1}{4},$$

$$|I_1(s)| \leq \int_N^\infty \frac{1}{x^2}\,dx = \frac{1}{N} \leq \frac{1}{2}.$$

Also,

$$|I_2(s)| \leq \int_N^\infty \frac{\log x}{x^{\sigma+1}}\,dx = \frac{\log N}{\sigma N^\sigma} + \frac{1}{\sigma^2 N^\sigma},$$

hence

$$|sI_2(s)| \leq \left(1 + \frac{t}{\sigma}\right)\frac{\log N + 1}{N} \leq \frac{1+t}{N}(\log N + 1) \leq 2(\log t + 1),$$

since $t + 1 \leq N + 2 \leq 2N$. Putting all the terms together, we have

$$|\zeta'(s)| \leq \tfrac{1}{2}(\log t)^2 + 2\log t + 4 < \tfrac{1}{2}(\log t + 3)^2.$$

□

The method of 3.1.6 can easily be adapted for other values of σ, for example, to show that $|\zeta(\frac{1}{2}+it)| = O(t^{1/2})$ (see exercise 6). However, it is clear that this method fails to take advantage of the cancellation occurring in sums like $\sum_{n=1}^{N} 1/n^s$. Better estimates are obtained through a more careful study of "exponential sums" of the form $\sum_{n=M+1}^{N} e^{if(n)}$ (this topic is discussed in many books; for a particularly straightforward account, see [GK]). Actually, the estimate of $|\zeta(1+it)|$ cannot be improved very much, but much stronger estimates can be established for other values of σ, notably

$$|\zeta(\tfrac{1}{2}+it)| = O(t^{1/6}\log t), \qquad |\zeta(it)| = O(t^{1/2}\log t).$$

The case $\sigma = \frac{1}{2}$ is of particular interest. With considerable effort, it has been shown that $|\zeta(\frac{1}{2}+it)|$ is $O(t^{\alpha})$ with α slightly less than $\frac{1}{6}$, but the best possible value of α is not known; indeed, the *Lindelöf hypothesis* is the conjecture (still unsolved) that $|\zeta(\frac{1}{2}+it)|$ is $o(t^{\epsilon})$ for *every* $\epsilon > 0$.

The proof that $\zeta(s) \neq 0$ *when* Re $s = 1$

Recall that the Euler product shows that $\zeta(s) \neq 0$ when Re $s > 1$. Our next objective is to show that this remains true on the line Re $s = 1$ (and on at least a narrow region to the left of this line). We will also make this more precise by giving an estimate for $1/|\zeta(1+it)|$ in terms of $\log t$. This is a crucial step in our journey towards the prime number theorem, because it ensures that the function $\zeta'(s)/\zeta(s)$ exists on and near this line. For the first statement, $\zeta(1+it) \neq 0$, we follow the simple and elegant method developed by de la Vallée Poussin and Mertens. It makes no reference to the actual formula used to extend the zeta function: it only depends on the fact that a differentiable extension can be defined somehow.

Lemma 3.1.9 *For any* θ, *we have* $3 + 4\cos\theta + \cos 2\theta \geq 0$.

Proof This follows from the identity

$$3 + 4\cos\theta + \cos 2\theta = 2 + 4\cos\theta + 2\cos^2\theta = 2(1+\cos\theta)^2. \qquad \square$$

Proposition 3.1.10 *Suppose that* $a(n) \geq 0$ *for all* n *and that the Dirichlet series* $\sum_{n=1}^{\infty} a(n)/n^s$ *converges, say to* $f(s)$, *when* Re $s > \sigma_0$. *Then, for* $\sigma > \sigma_0$,

$$3f(\sigma) + 4\text{Re } f(\sigma+it) + \text{Re } f(\sigma+2it) \geq 0.$$

Proof We have

$$3f(\sigma) + 4f(\sigma + it) + f(\sigma + 2it) = \sum_{n=1}^{\infty} \frac{a(n)}{n^{\sigma}}(3 + 4n^{-it} + n^{-2it}).$$

But Re $(3 + 4n^{-it} + n^{-2it}) = 3 + 4\cos\theta_n + \cos 2\theta_n$, where $\theta_n = t \log n$. The statement follows, by the previous lemma. □

Corollary 3.1.11 *For all $\sigma > 1$ and all t, we have*

$$\zeta(\sigma)^3 |\zeta(\sigma + it)|^4 |\zeta(\sigma + 2it)| \geq 1.$$

Proof Apply 3.1.10 to $\log \zeta(s)$. By 2.3.2, this function is expressible, for Re $s > 1$, in the form $\sum_{n=1}^{\infty} a(n)/n^s$, with $a(n) \geq 0$. Since Re $\log z = \log |z|$, the statement follows. □

Theorem 3.1.12 *For all $t \neq 0$, we have $\zeta(1 + it) \neq 0$.*

Proof Suppose that $\zeta(1 + it) = 0$ for some $t \neq 0$. Then

$$\frac{\zeta(\sigma + it)}{\sigma - 1} \to \zeta'(1 + it) \quad \text{as } \sigma \to 1^+.$$

Let $H(\sigma)$ denote the expression in 3.1.11. Then

$$H(\sigma) = [(\sigma - 1)\zeta(\sigma)]^3 \left(\frac{|\zeta(\sigma + it)|}{\sigma - 1} \right)^4 (\sigma - 1)|\zeta(\sigma + 2it)|.$$

But by 1.7.1, we know that $(\sigma - 1)\zeta(\sigma) \to 1$ as $\sigma \to 1^+$, while $\zeta(\sigma + 2it)$ tends to $\zeta(1 + 2it)$. Hence $H(\sigma) \to 0$ as $\sigma \to 1^+$, contradicting 3.1.11. □

Note that this proof worked because the coefficient 4 in 3.1.9 is bigger than the constant term 3, giving a spare power of $(\sigma - 1)$.

Purely by continuity, we can deduce that the zero-free region extends a bit to the left of the line. We use the following result from elementary analysis: *every bounded real sequence has a convergent subsequence.*

Proposition 3.1.13 *Let $0 < a < b$. Then there exists $\delta > 0$ such that $\zeta(\sigma + it) \neq 0$ when $1 - \delta \leq \sigma \leq 1$ and $a \leq t \leq b$.*

Proof Suppose that the statement is false. Then, for each $n \geq 1$, there exists $s_n = \sigma_n + it_n$ such that $1 - \frac{1}{n} \leq \sigma_n \leq 1$, $a \leq t_n \leq b$ and $\zeta(s_n) = 0$. The sequence (t_n) has a subsequence (t_{n_j}) that converges, say to t_0. Let $s_{n_j} = \sigma_{n_j} + it_{n_j}$. Then $\lim_{j \to \infty} s_{n_j} = 1 + it_0$. Since ζ is continuous, it follows that $\zeta(1 + it_0) = 0$, contrary to 3.1.12. □

We now come to the quantitative version (this result, like 3.1.8, is not needed for Newman's method). We use the same lemmas again, together with the estimates for $|\zeta(s)|$ and $|\zeta'(s)|$ in terms of t.

Theorem 3.1.14 *Suppose that, for all $\sigma \geq 1$ and $t \geq t_0$, we have*

$$|\zeta(\sigma + 2it)| \leq M_1(t), \qquad |\zeta'(\sigma + it)| \leq M_2(t),$$

where $M_1(t), M_2(t) \geq 1$. Then, for such σ and t,

$$\frac{1}{|\zeta(\sigma + it)|} \leq 2^5 M_1(t) M_2(t)^3.$$

Hence, in particular, for $t \geq 2$ and $\sigma \geq 1$, we have

$$\frac{1}{|\zeta(\sigma + it)|} \leq 4(\log t + 5)^7.$$

Proof We prove the statement for $\sigma > 1$: the result for $\sigma = 1$ then follows by continuity. Since

$$\frac{1}{|\zeta(s)|} \leq \zeta(\sigma) \leq \frac{\sigma}{\sigma - 1},$$

we have $1/|\zeta(s)| \leq 5$ whenever $\sigma > \frac{5}{4}$: clearly, the statment then holds for all $t > 1$. So we assume that $\sigma \leq \frac{5}{4}$. Note that we then have $\zeta(\sigma) \leq 2^{1/3}/(\sigma - 1)$, since $\frac{5}{4} < 2^{1/3}$.

We write M_1, M_2 for $M_1(t)$ and $M_2(t)$. For any σ in $(1, \frac{5}{4})$, we have by 3.1.11 and the estimate just given for $\zeta(\sigma)$

$$\frac{2}{(\sigma - 1)^3} |\zeta(\sigma + it)|^4 M_1 \geq 1,$$

or $|\zeta(\sigma + it)| \geq f(\sigma)$, where

$$f(\sigma) = \frac{(\sigma - 1)^{3/4}}{2^{1/4} M_1^{1/4}}.$$

Now define the number η by the equation

$$f(\eta) = 2M_2(\eta - 1).$$

From the expression for $f(\eta)$, this means that

$$\eta - 1 = \frac{1}{2^5 M_1 M_2^4}.$$

(Note that certainly $\eta - 1 \leq \frac{1}{4}$.) Suppose first that $1 < \sigma < \eta$. Since

$$\zeta(\eta + it) - \zeta(\sigma + it) = \int_\sigma^\eta \zeta'(x + it)\, dx,$$

we have

$$|\zeta(\eta + it) - \zeta(\sigma + it)| \le M_2(\eta - \sigma) < M_2(\eta - 1). \tag{3.7}$$

Now $|\zeta(\eta + it)| \ge f(\eta) = 2M_2(\eta - 1)$, so it follows from (3.7) that

$$|\zeta(\sigma + it)| \ge M_2(\eta - 1) = \frac{1}{2^5 M_1 M_2^3}.$$

Now suppose instead that $\eta \le \sigma \le \frac{5}{4}$. Then we have

$$|\zeta(\sigma + it)| \ge f(\sigma) \ge f(\eta) = \frac{1}{2^4 M_1 M_2^3}.$$

In both cases, the stated inequality holds.

Finally, by 3.1.6 and 3.1.8, we can take $M_1(t) = \log 2t + 4 < \log t + 5$ and $M_2(t) = \frac{1}{2}(\log t + 3)^2$. □

Again, the $\log t$ factor cannot be removed: it can be shown that $1/|\zeta(1+it)|$ is unbounded for $t > 1$, or, in other words, there are values of t for which $\zeta(1 + it)$ comes arbitrarily close to 0 (which shows that 3.1.12 was by no means a foregone conclusion!).

A much stronger result is suspected. The zeta function is known to have zeros (in fact, infinitely many) on the line Re $s = \frac{1}{2}$. The *Riemann hypothesis* is the name given to the conjecture that it has *no* zeros in the region Re $s > \frac{1}{2}$. This was suggested by Riemann in his pioneering work of 1859. After more than 140 years, it remains one of the most celebrated unsolved problems in mathematics – in the view of many mathematicians, the most important one of all. As we shall see in section 5.2, the hypothesis is equivalent to a much stronger estimate than the known ones for the accuracy of the approximation to $\pi(x)$ by $\text{li}(x)$. It would also imply the Lindelöf hypothesis.

Methods have been devised for precise computation of zeros of the zeta function in order of increasing distance from the real axis. Massive computational efforts by Odlyzko, te Riele and others have shown that the first 5×10^9 zeros all lie on the line Re $s = \frac{1}{2}$; this means all the zeros with $0 < t \le t_0$, where t_0 is 1.7×10^9. Further computations are increasing these numbers all the time. This is clearly compelling numerical evidence in favour of the Riemann hypothesis. However, it is no more than that, and it does not even begin to point the way to a proof of the general statement. It is not even known whether there is any value $\alpha < 1$ such that all zeros satisfy $\sigma \le \alpha$. It is only known that, for a certain $c > 0$, there are no zeros in the region $\sigma > 1 - c/\log t$. We give a proof of this in section 5.3, while at the same time showing that the unlikely looking power 7 in 3.1.14 is not really needed.

Sketch of $\zeta(1+it)$ *for* $0 < t \leq 36$

We pause here for a sketch. Of course, it is impossible to give a satisfactory representation of a complex-valued function in a single sketch. But it can be quite illuminating to sketch the curve in the complex plane described by $f(s)$ when s moves along a chosen line. For the zeta function, an interesting choice is the vertical line $s = 1 + it$. The sketch shows the curve traced out by $\zeta(1 + it)$ for $0 < t \leq 36$. It comprises a succession of clockwise loops of varying sizes, with none of the regularity displayed by the more familiar "standard" functions.

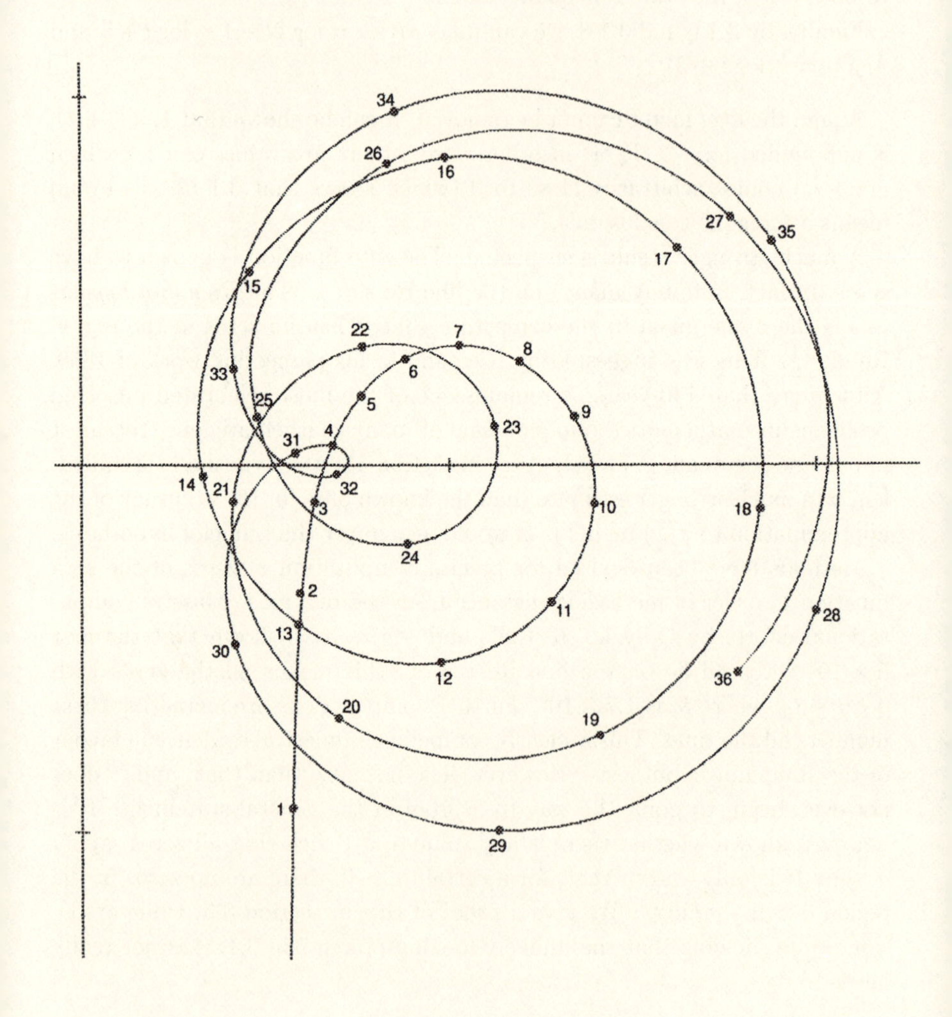

Extension to Re $s > -1$

If we use $x - [x] - \frac{1}{2}$ instead of $x - [x]$, we obtain better results at the cost of slightly more complicated expressions. (This subsection is included for interest, but it is not needed for either version of the proof of the prime number theorem.)

To simplify the formulae a bit, we shall write $x-[x]-\frac{1}{2} = B(x)$. Formulae (3.2) and (3.3) show how $\zeta(s)$ is expressed using $B(x)$.

Lemma 3.1.15 *For any $s = \sigma + it$ with $\sigma > -1$, the integral*

$$\int_1^\infty \frac{B(x)}{x^{s+1}}\, dx$$

is convergent, say to $I^*(s)$. Further, if

$$I_N^*(s) = \int_1^N \frac{B(x)}{x^{s+1}}\, dx,$$

then

$$|I^*(s) - I_N^*(s)| \le \frac{|s+1|}{\sigma+1}\, \frac{1}{8N^{\sigma+1}}.$$

Hence $I_N^*(s) \to I^*(s)$ as $N \to \infty$, uniformly on every set of the form $\{s : \sigma \ge -1 + \delta,\ |t| \le T\}$.

Proof In 1.4.12, take $f(x) = 1/x^s$ and remove the factor s: we obtain

$$\begin{aligned}\left|\int_N^R \frac{B(x)}{x^{s+1}}\, dx\right| &\le \tfrac{1}{8}\int_N^R \frac{|s+1|}{|x^{s+2}|}\, dx \\ &= \tfrac{1}{8}|s+1| \int_N^R \frac{1}{x^{\sigma+2}}\, dx \\ &\le \frac{|s+1|}{\sigma+1}\, \frac{1}{8N^{\sigma+1}}.\end{aligned}$$

By the Cauchy principle for convergence, it follows that the stated integral for $I^*(s)$ is convergent. Also, the stated inequality follows by taking the limit as $R \to \infty$. Now fix T and δ, and let $s = \sigma + it$ belong to the set described. Then

$$\left|\frac{s+1}{\sigma+1}\right| = \left|1 + i\frac{t}{\sigma+1}\right| \le 1 + \frac{T}{\delta},$$

so

$$|I^*(s) - I_N^*(s)| \le \left(1 + \frac{T}{\delta}\right) \frac{1}{8N^\delta},$$

which tends to 0 as $N \to \infty$. This proves uniform convergence. □

Hence we can use (3.2) and (3.3) to extend the definition of $\zeta(s)$ to $\operatorname{Re} s > -1$, and the extended function is again holomorphic. We derive the statement corresponding to 3.1.3. By Euler's summation formula in the form 1.4.7,

$$\sum_{n=1}^{N-1} \frac{1}{n^s} + \frac{1}{2N^s} = \frac{1}{2} + \frac{1}{s-1} - \frac{N^{1-s}}{s-1} - s\int_1^N \frac{B(x)}{x^{s+1}}\,dx. \tag{3.8}$$

Proposition 3.1.16 *The function $\zeta(s)$, defined by (3.2) and (3.3), is defined and holomorphic for all $s \neq 1$ with* $\operatorname{Re} s > -1$. *For all integers $N \geq 1$,*

$$\zeta(s) = \sum_{n=1}^{N-1} \frac{1}{n^s} + \frac{1}{2N^s} + \frac{N^{1-s}}{s-1} - r_N^*(s), \tag{3.9}$$

where

$$r_N^*(s) = s\int_N^\infty \frac{B(x)}{x^{s+1}}\,dx, \tag{3.10}$$

and we have

$$|r_N^*(s)| \leq \frac{|s(s+1)|}{\sigma+1}\,\frac{1}{8N^{\sigma+1}}.$$

Proof Formula (3.9) is obtained by subtracting (3.8) from (3.2), and the other statements follow from the previous lemma. □

From the fact that $|B(x)| \leq \frac{1}{2}$, we also have the more elementary inequality $|r_N^*(s)| \leq |s|/(2\sigma N^\sigma)$ for $\sigma > 0$.

The expression in (3.9) opens the way to computation of $\zeta(s)$ with reasonable accuracy. Write the right-hand side of (3.9) as $\zeta_N^*(s) - r_N^*(s)$. Consider the problem of calculating $\zeta(1+i)$. By 3.1.16, the error in approximating to it by $\zeta_N^*(1+i)$ is no more than $\sqrt{10}/16N^2$. The comparable estimate if we approximated using (3.5) would be $\sqrt{2}/N$. As before, by halving the Nth term, we gain accuracy by a factor of $7N$.

With more effort, the zeta function can be extended so that it is defined on the whole complex plane (with the known pole at 1, but no other singularities). This is not important for our proof of the prime number theorem, but it is used in some alternative approaches. One way to perform the extension is by repeated integration by parts in (3.2): each step increases the region of convergence by a strip of width 1. The process is known as *Euler-Maclaurin summation*, and the resulting formulae involve the *Bernoulli* numbers and polynomials. See [Edw].

A more satisfactory method involves the complex gamma function $\Gamma(s)$.

One shows, by any of several methods, that the following identity holds when $-1 < \text{Re } s < 0$:

$$\zeta(s) = 2^s \pi^{s-1} \Gamma(1-s) \sin \tfrac{1}{2} s\pi \, \zeta(1-s).$$

Now the right-hand side defines a holomorphic function for all s with $\text{Re } s < 0$ (note that then $\text{Re } (1-s) > 1$, so $\zeta(1-s)$ is already defined). Hence we can use this formula to define $\zeta(s)$ for such s. It is known as the *functional equation for the zeta function* and can be written in a number of equivalent forms. See [Ell], [Ap], [Ivić], [Titch] or [Patt].

Exercises

1 Show directly from the defining formula for $\zeta(s)$ that, for $0 < \sigma < 1$,

$$-\frac{1}{1-\sigma} \le \zeta(\sigma) \le -\frac{\sigma}{1-\sigma}.$$

2 Show that if $0 < \text{Re } s < 1$, then

$$\zeta(s) = -s \int_0^\infty \frac{x - [x]}{x^{s+1}} \, dx.$$

3 Use formula (3.5) with N and $2N$ to show that, for all $s \neq 1$ with $\text{Re } s > 0$,

$$\sum_{n=1}^\infty \frac{(-1)^{n-1}}{n^s} = (1 - 2^{1-s})\zeta(s).$$

4 Let $f(s) = 1/\zeta(s)$ for $s \neq 1$ and $f(1) = 0$. What are the values of $f'(1)$ and $f''(1)$? Deduce the value of $\lim_{s \to 1} \zeta'(s)/\zeta(s)^2$.

5 Modify the proof of 3.1.6 to show that if $t \ge 2$ and $\sigma \ge 1 - a/\log t$ (and also $\sigma \ge \frac{1}{2}$), then

$$|\zeta(\sigma + it)| \le e^a(\log t + 5).$$

(Note that if $n \le t$, then $n^{1-\sigma} \le e^a$, so that $n^{-\sigma} \le e^a/n$.)

6 Prove that $|\zeta(\frac{1}{2} + it)| \le 4t^{1/2} + 1$ for all $t \ge 4$.

7 (*A series expression for the extended zeta function*) Use the second expression for $\zeta(s)$ in 3.1.3 to show that, for $s \neq 1$ with $0 < \text{Re } s \le 1$,

$$\zeta(s) = \sum_{n=1}^\infty \left(\frac{1}{n^s} - \frac{1}{1-s}\left(n^{1-s} - (n-1)^{1-s}\right) \right),$$

in which 0^{1-s} is equated to 0.

8 Use 3.1.2 to show that if $\zeta(s) = 0$, where $s = \sigma + it$, then $|(s+1)/(s-1)| \leq |s|/\sigma$. Deduce that ζ has no zeros in the region $\frac{3}{4} \leq \sigma \leq 1$, $|t| \leq 1$. (Show that $|s|/\sigma \leq \frac{5}{3}$ in this region.)

9 Use 2.2.4 to show that if $\zeta(s_0) = 0$, where $s_0 = 1+it_0$, then $\zeta'(s_0) \neq 0$. Now give an alternative proof of 3.1.12 by applying 3.1.10 to the series for $\zeta'(s)/\zeta(s)$. (Note that if $f(s)$ has a simple zero at s_0, then $f'(s)/f(s) = 1/(s-s_0) + g(s)$, where g is holomorphic at s_0.)

10 For real $a > 0$, the "Hurwitz zeta function" is defined by

$$\zeta(s,a) = \sum_{n=0}^{\infty} \frac{1}{(n+a)^s}$$

for Re $s > 1$ (so that $\zeta(s,1) = \zeta(s)$). Write down formulae extending $\zeta(s,a)$ to Re $s > 0$ and Re $s > -1$. Show that the extended definition gives $\zeta(0,a) = \frac{1}{2} - a$.

11 The real and imaginary parts of $\zeta(1+i)$ are known to be 0.58216 and -0.92685 to five decimal places. Find the approximations to these numbers (to four decimal places) given by formula (3.9), with $N = 10$.

Optional: Estimate some more, e.g. $\zeta(\frac{3}{2})$, $\zeta(2+i)$, $\zeta(1+2i)$.

3.2 Inversion of Dirichlet series; the integral version of the fundamental theorem

Inversion of Dirichlet series by integrals on vertical lines

Let $a(n)$ be an arithmetic function, and let $f(s)$ be the function defined by its corresponding Dirichlet series:

$$f(s) = \sum_{n=1}^{\infty} \frac{a(n)}{n^s}.$$

As usual, write $A(x) = \sum_{n \leq x} a(n)$. Our aim is to deduce an estimation of $A(x)$ from information about $f(s)$. In the present section, we shall carry out this programme as far as what we shall call the "integral version of the fundamental theorem". Once this is known, it is a fairly short step to derive the "limit" and "series" versions, incorporating the prime number theorem as a particular case. These versions are presented in section 3.4.

An interesting alternative route to the integral version (under slightly different conditions) is described in section 3.3. However, the method of the present section is recommended to readers intending to go on to chapter 5.

The first stage, roughly, is to "invert" the series to express $A(x)$ in terms of $f(s)$. Actually, it is easier to find an expression for

$$\int_1^x \frac{A(y)}{y^2}\, dy$$

instead of $A(x)$ itself; this will serve just as well for our ultimate purpose. The expression will be an integral involving $f(s)$, necessarily along some path in the complex plane. Note that the corresponding step for *power* series is very simple: if $g(s) = \sum_{n=0}^{\infty} a(n)s^n$ for $|s| < R$, then $2\pi i a(n)$ equals the integral of $g(s)/s^{n+1}$ round a circle of radius less than R. Given that a Dirichlet series converges on a half-plane, it is not surprising that the right kind of path for such a series is a vertical line.

This is the point at which complex integration enters our strategy. We will only need the really basic theorems of the subject, Cauchy's integral theorem and formula, together with the following fact: if $\ell(\Gamma)$ denotes the length of a path Γ in the complex plane, and $|f(s)| \le M$ for s on Γ, then $|\int_\Gamma f(s)\, ds| \le M\ell(\Gamma)$.

The notation $\int_{c-iT}^{c+iT} f(s)\, ds$ (where c, T are real) means the integral of $f(s)$ along the straight line from $c - iT$ to $c + iT$, in other words,

$$\int_{-T}^{T} f(c+it)i\, dt.$$

Provided that the integrals of this function on $[-T, 0]$ and $[0, T]$ tend to limits separately as $T \to \infty$, the limit as $T \to \infty$ of the above integral is written as $\int_{c-i\infty}^{c+i\infty} f(s)\, ds$. For convenience, we shall replace this notation by $\int_{L_c} f(s)\, ds$.

We also write

$$E(x) = \begin{cases} 1 & \text{if } x \ge 1, \\ 0 & \text{if } x < 1. \end{cases}$$

Our starting point is the following integral. Note that x is fixed and integration is with respect to s.

Proposition 3.2.1 *If $x > 0$ and $c > 0$, then*

$$\frac{1}{2\pi i}\int_{L_c} \frac{x^s}{s^2}\, ds = E(x)\log x.$$

Proof Let C be the circle $C(0, R)$, where $R > c$. Let the line $\operatorname{Re} s = c$ meet this circle at $c \pm it_R$. Let L_R be the line segment from $c - it_R$ to $c + it_R$, and let C_1, C_2 (respectively) be the sections of C to the left and right of this line.

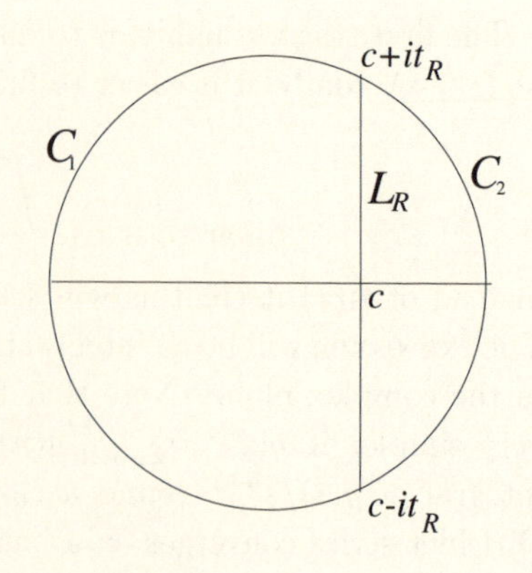

For any path Γ in the complex plane, write

$$I(\Gamma) = \frac{1}{2\pi i} \int_\Gamma \frac{x^s}{s^2}\, ds.$$

Now $x^s = e^{\lambda s}$, where $\lambda = \log x$, hence

$$\frac{x^s}{s^2} = \frac{e^{\lambda s}}{s^2} = \frac{1}{s^2}(1 + \lambda s + \tfrac{1}{2}\lambda^2 s^2 + \cdots),$$

and this series is uniformly convergent on any set of the form $r_1 \le |s| \le r_2$. Hence the series can be integrated termwise on $L_R \cup C_1$, which is a closed contour enclosing the point 0. It is elementary that the integral of s^{-1} around this contour is $2\pi i$, while the integral of all other powers s^n is 0. So we have

$$I(L_R \cup C_1) = \lambda.$$

Now $|x^s| = x^\sigma$, where $s = \sigma + it$. So, if $x \ge 1$, then for s on C_1 we have $|x^s| \le x^c$, so that $|x^s/s^2| \le x^c/R^2$ and hence

$$|I(C_1)| \le \frac{1}{2\pi} \frac{x^c}{R^2} 2\pi R = \frac{x^c}{R},$$

which tends to 0 as $R \to \infty$. Hence

$$I(L_R) \to \lambda = \log x \quad \text{as } R \to \infty.$$

This is our statement for $x \ge 1$ (note that separate convergence on $[0, t_R]$ and $[-t_R, 0]$ is clear, since $|x^s/s^2| \le x^c/t^2$).

If $0 < x < 1$, we consider instead $I(C_2 \cup L_R)$. The integrand x^s/s^2 has no poles inside this contour, so, by Cauchy's theorem, the integral equals 0. The function x^σ now decreases with x, so $|x^s| \le x^c$ for s on C_2. In the same way as before, $I(C_2) \to 0$ as $R \to \infty$, and hence $I(L_R) \to 0$. □

Note It follows that if $c' > 1$, then

$$\frac{1}{2\pi i} \int_{L_{c'}} \frac{x^{s-1}}{(s-1)^2}\, ds = E(x) \log x.$$

In fact, by writing $c' = c + 1$ and substituting $s = c' + it$, we see that this integral is identical to the one in 3.2.1.

Proposition 3.2.2 *If $x > 0$ and $c > 1$, then*

$$\frac{1}{2\pi i}\int_{L_c} \frac{x^s}{s(s-1)}\,ds = (x-1)E(x).$$

Proof This is similar to 3.2.1. To show that the integrals on the circular arcs tend to 0, we use $|s(s-1)| \geq R(R-1)$ on C (instead of $|s^2| = R^2$). To evaluate $I(L_R \cup C_1)$, note that

$$\frac{x^s}{s(s-1)} = \frac{x^s}{s-1} - \frac{x^s}{s}.$$

By Cauchy's integral formula, it follows that

$$I(L_R \cup C_1) = x^1 - x^0 = x - 1.$$

The statements follow, as before. □

Now suppose that $f(s) = \sum_{n=1}^{\infty} a(n)/n^s$ for Re $s > 1$. If termwise integration can be justified, then 3.2.2 will give

$$\begin{aligned}\frac{1}{2\pi i}\int_{L_c} \frac{x^s}{s(s-1)} f(s)\,ds &= \sum_{n=1}^{\infty} a(n)\frac{1}{2\pi i}\int_{L_c} \frac{1}{s(s-1)}\left(\frac{x}{n}\right)^s ds \\ &= \sum_{n=1}^{\infty} a(n)E\left(\frac{x}{n}\right)\left(\frac{x}{n}-1\right) \\ &= \sum_{n\leq x} a(n)\left(\frac{x}{n}-1\right),\end{aligned}$$

since $E(x/n) = 0$ when $n > x$. The next theorem says that this is true (but for future use, the statement is given with both sides divided by x).

Theorem 3.2.3 *Suppose that the Dirichlet series $\sum_{n=1}^{\infty} a(n)/n^s$ is absolutely convergent for* Re $s > 1$*, with sum $f(s)$. Let $A(x) = \sum_{n\leq x} a(n)$. Then, for $c > 1$ and $x > 1$,*

$$\frac{1}{2\pi i}\int_{L_c} \frac{x^{s-1}}{s(s-1)} f(s)\,ds = \sum_{n\leq x} a(n)\left(\frac{1}{n} - \frac{1}{x}\right).$$

By Abel's summation formula, this can also be written

$$\int_1^x \frac{A(y)}{y^2}\,dy.$$

Proof Write $x^s f(s) = G(s) + H(s)$, where

$$G(s) = \sum_{n\leq x} a(n)\left(\frac{x}{n}\right)^s, \qquad H(s) = \sum_{n>x} a(n)\left(\frac{x}{n}\right)^s.$$

For $G(s)$ (which is only a finite sum) we have as above

$$\frac{1}{2\pi i}\int_{L_c} \frac{G(s)}{s(s-1)}\,ds = \sum_{n\le x} a(n)\left(\frac{x}{n}-1\right).$$

Now $\sum_{n>x} |a(n)|(x/n)^c$ is convergent, say to M. For $n > x$ and Re $s \ge c$, we have $|(x/n)^s| \le (x/n)^c$, and hence $|H(s)| \le M$. Now consider

$$\int_{C_2\cup L_R} \frac{H(s)}{s(s-1)}\,ds$$

as in the proof of 3.2.1. Since $H(s)$ is differentiable for Re $s > 1$, this integral equals 0, by Cauchy's theorem, and exactly as before we see that the contribution of C_2 tends to 0 as $R \to \infty$. Hence

$$\frac{1}{2\pi i}\int_{L_c} \frac{H(s)}{s(s-1)}\,ds = 0.$$

To obtain the first statement, now divide both sides by x. The second statement follows, by Abel's summation formula in the form 1.3.6(ii), applied to the function $1/y$. □

The Riemann-Lebesgue lemma

The following result, the "Riemann-Lebesgue lemma", is needed for either proof of the fundamental theorems. Let ϕ be a differentiable, complex-valued function on $\mathbb{R}$ such that $\int_{-\infty}^{\infty} |\phi(t)|\,dt$ is convergent. For such a function, we can define (for all real λ)

$$F(\lambda) = \int_{-\infty}^{\infty} e^{i\lambda t}\phi(t)\,dt.$$

This integral is convergent, since $|e^{i\lambda t}\phi(t)| = |\phi(t)|$. (The function $F(-\lambda)$ is the *Fourier transform* of ϕ.)

Proposition 3.2.4 *Let ϕ be a complex-valued function on $\mathbb{R}$ with continuous derivative and such that $\int_{-\infty}^{\infty} |\phi(t)|\,dt$ is convergent. Let $F(\lambda)$ be as above. Then $F(\lambda) \to 0$ as $\lambda \to \infty$.*

Proof Take $\varepsilon > 0$. There exists T such that $\int_T^{\infty} |\phi(t)|\,dt \le \varepsilon$ (and similarly for $(-\infty, -T]$). For any λ, we then have

$$\left|\int_T^{\infty} e^{i\lambda t}\phi(t)\,dt\right| \le \varepsilon.$$

Now $|\phi'(t)|$ is bounded (since continuous), on $[-T, T]$, say by M. Let

$$F_T(\lambda) = \int_{-T}^{T} e^{i\lambda t}\phi(t)\, dt.$$

Integrating by parts, we have

$$F_T(\lambda) = \frac{1}{i\lambda}\left[e^{i\lambda t}\phi(t)\right]_{-T}^{T} - \frac{1}{i\lambda}\int_{-T}^{T} e^{i\lambda t}\phi'(t)\, dt,$$

and hence

$$|F_T(\lambda)| \le \frac{1}{\lambda}\left[|\phi(T)| + |\phi(-T)|\right] + \frac{2MT}{\lambda}.$$

So $F_T(\lambda) \to 0$ as $\lambda \to \infty$, and for sufficiently large λ we have $|F_T(\lambda)| \le \varepsilon$. Adding the contributions of $[T, \infty)$ and $(-\infty, -T]$, we see that $|F(\lambda)| \le 3\varepsilon$ for such λ. □

Note 1 Of course, we have proved the same conclusion for $F_T(\lambda)$.

Note 2 If we know that $\int_{-\infty}^{\infty} |\phi'(t)|\, dt$ is also convergent (with value I_1, say) and $\phi(\pm t) \to 0$ as $t \to \infty$, then integration by parts as above gives

$$F(\lambda) = -\frac{1}{i\lambda}\int_{-\infty}^{\infty} e^{i\lambda t}\phi'(t)\, dt,$$

and hence $|F(\lambda)| \le I_1/|\lambda|$. This has the advantage of giving an estimate of the *rate* of convergence to 0. We return to this point in chapter 5.

The integral version of the fundamental theorem

We are now ready to state this theorem. For present purposes, the word "region" simply means a subset of the complex plane.

Theorem 3.2.5 *Suppose that f is a complex function differentiable on a region including* Re $s \ge 1$, *except possibly at the point 1, and that:*

(FT1) *the series* $\displaystyle\sum_{n=1}^{\infty} \frac{a(n)}{n^s}$ *converges absolutely to $f(s)$ when* Re $s > 1$;

(FT2) $f(s) = \dfrac{\alpha}{s-1} + \alpha_0 + (s-1)h(s)$, *where h is differentiable at 1;*

(FT3) *there is a function $P(t)$ such that $|f(\sigma \pm it)| \le P(t)$ when $\sigma \ge 1$ and $t \ge t_0$ (where $t_0 \ge 1$), and also* $\displaystyle\int_1^{\infty} \frac{P(t)}{t^2}\, dt$ *is convergent.*

Then $\displaystyle\int_1^{\infty} \frac{A(x) - \alpha x}{x^2}\, dx$ *converges to* $\alpha_0 - \alpha$.

Proof Write $\phi(s) = h(s)/s$. Note that $h(s)$ and $\phi(s)$ are differentiable at all points s (including 1) with $\operatorname{Re} s \geq 1$. Using the identity $1/(s-1) = s/(s-1) - 1$, we have

$$(s-1)h(s) = f(s) - \frac{\alpha}{s-1} - \alpha_0 = f(s) - \alpha\frac{s}{s-1} - \alpha',$$

where $\alpha' = \alpha_0 - \alpha$, and hence

$$\phi(s) = \frac{f(s)}{s(s-1)} - \frac{\alpha}{(s-1)^2} - \frac{\alpha'}{s(s-1)}. \tag{3.11}$$

For all $s = \sigma + it$ with $\sigma \geq 1$ and $|t| \geq t_0$, we have $|s(s-1)\phi(s)| \leq P_1(t)$, where $P_1(t) = P(t) + |\alpha| + |\alpha_0|$. For such s, we have $|s(s-1)| \geq t^2$, so that

$$|\phi(s)| \leq \frac{P_1(t)}{t^2}. \tag{3.12}$$

Of course, $\int_1^\infty [P_1(t)/t^2]\, dt$ is convergent.

For $x > 1$ and $c \geq 1$, define

$$I(x,c) = \frac{1}{2\pi i}\int_{L_c} x^{s-1}\phi(s)\, ds.$$

For any $c > 1$, we have by (3.11) and 3.2.1, 3.2.2 and 3.2.3:

$$\begin{aligned} I(x,c) &= \frac{1}{2\pi i}\int_{L_c} \frac{x^{s-1}}{s(s-1)} f(s)\, ds - \frac{\alpha}{2\pi i}\int_{L_c} \frac{x^{s-1}}{(s-1)^2}\, ds \\ &\quad - \frac{\alpha'}{2\pi i}\int_{L_c} \frac{x^{s-1}}{s(s-1)}\, ds \\ &= \int_1^x \frac{A(y)}{y^2}\, dy - \alpha \log x - \alpha'\left(1 - \frac{1}{x}\right) \\ &= \int_1^x \frac{A(y) - \alpha y}{y^2}\, dy - \alpha'\left(1 - \frac{1}{x}\right). \end{aligned}$$

Note that this is independent of $c > 1$. We will show that $I(x,1)$ has the same value. Suppose for the moment that this is correct. Then, writing $s = 1 + it$, we have

$$I(x,1) = \frac{1}{2\pi}\int_{-\infty}^{\infty} x^{it}\phi(1+it)\, dt = \frac{1}{2\pi}\int_{-\infty}^{\infty} e^{i\lambda t}\phi(1+it)\, dt,$$

where $\lambda = \log x$. By (3.12), $\int_{-\infty}^{\infty} |\phi(1+it)|\, dt$ is convergent. It follows from 3.2.4 that $I(x,1) \to 0$ as $x \to \infty$. In other words,

$$\int_1^x \frac{A(y) - \alpha y}{y^2}\, dy \to \alpha' \quad \text{as } x \to \infty.$$

This is our statement.

It remains to justify the statement that $I(x,1) = I(x,c)$ for $c > 1$. Now

$$I(x,c) = \frac{1}{2\pi}\int_{-\infty}^{\infty} g(c,t)\,dt,$$

where

$$g(c,t) = x^{c-1+it}\phi(c+it).$$

Choose $\varepsilon > 0$. There exists $T \geq t_0$ such that

$$\int_T^{\infty} \frac{P_1(t)}{t^2}\,dt \leq \varepsilon.$$

If $1 \leq c \leq 2$, then $|x^{c-1+it}| = x^{c-1} \leq x$, so, by (3.12),

$$\int_T^{\infty} |g(c,t)|\,dt \leq \varepsilon x.$$

A similar estimate applies to the integral on $(-\infty, -T]$. Finally, since $g(c,t)$ is continuous on the closed rectangle $1 \leq c \leq 2$, $-T \leq t \leq T$, it is *uniformly* continuous there, so that if c is close enough to 1, then $|g(c,t) - g(1,t)| \leq \varepsilon/2T$ for all t in $[-T,T]$. For such c,

$$\left| \int_{-T}^{T} g(c,t)\,dt - \int_{-T}^{T} g(1,t)\,dt \right| \leq \varepsilon.$$

Together with the contributions from outside $[-T,T]$, this gives

$$|I(x,c) - I(x,1)| \leq \varepsilon(1+2x).$$

Since $I(x,c)$ is independent of $c > 1$, this shows that $I(x,1) = I(x,c)$, as required.

(An alternative method for the last part is to apply Cauchy's integral theorem on the rectangle with vertices $1 \pm iT$, $c \pm iT$ and show that the contribution of the horizontal sides tends to 0 as $T \to \infty$.) □

Note It was essential to move the line of integration to L_1, since otherwise the integral would contain x^{c-1} as a factor, destroying its chances of tending to 0 as $x \to \infty$. If, in a particular case, one could move the line to L_c where $c < 1$, then of course the factor x^{c-1} would *accelerate* the convergence to 0.

Example In advance of more interesting examples, let us see what the theorem says in the simplest case, $a(n) = 1$ and $f(s) = \zeta(s)$. Then $A(x) = [x]$ and $\alpha = 1$, $\alpha_0 = \gamma$, so the statement is

$$\int_1^{\infty} \frac{[x] - x}{x^2}\,dx = \gamma - 1,$$

which of course we know from 1.4.11.

We defer further examples of 3.2.5 until section 3.4, where they will be given alongside corresponding statements for limits and series.

Further remarks

Two variants of 3.2.3, proved by easy modifications of the same method, are

$$\frac{1}{2\pi i}\int_{L_c}\frac{x^s}{s^2}f(s)\,ds=\sum_{n\le x}a(n)(\log x-\log n)=\int_1^x\frac{A(y)}{y}\,dy, \tag{3.13}$$

$$\frac{1}{2\pi i}\int_{L_c}\frac{x^s}{s(s+1)}f(s)\,ds=\sum_{n\le x}a(n)\left(1-\frac{n}{x}\right)=\frac{1}{x}\int_1^x A(y)\,dy. \tag{3.14}$$

Many books use one of these formulae. If, for example, (3.13) is used, then the method of 3.2.5 leads to

$$\frac{1}{x}\int_1^x\frac{A(y)}{y}\,dy\to\alpha\quad\text{as } x\to\infty.$$

This is sufficient to give the *limit* version of the fundamental theorem later, but not the *series* version; hence our preference for the form chosen.

There is, in fact, a more direct inversion theorem along the lines first suggested: for non-integer values of x (avoiding discontinuities of $A(x)$),

$$\frac{1}{2\pi i}\int_{L_c}\frac{x^s}{s}f(s)\,ds=A(x). \tag{3.15}$$

This is called "Perron's inversion formula". It might be thought that it is a more obvious formula to use than any of the previous ones. However, with only s in the denominator, the previous proof fails, since we would be dividing by R instead of R^2. In fact, both the proof of (3.15) and its application demand careful evaluations of integrals on finite intervals $[c-iT,c+iT]$ instead of the whole of L_c, using rectangles instead of circles. See [Ap] or [Widd] (also exercise 4).

By Abel's summation formula,

$$f(s)=s\int_1^\infty\frac{A(x)}{x^{s+1}}\,dx,$$

so in all these statements both sides can be expressed in terms of $A(x)$, and one might suspect that the statements are valid for any suitable function $A(x)$, whether or not it is the summation function of an arithmetic function $a(n)$. This is, indeed, the case, and in fact all the statements are special cases of a more general result, the *Mellin inversion theorem.* Given a function $B(x)$ on $[0, \infty)$, let

$$g(s) = \int_0^\infty \frac{B(x)}{x^{s+1}}\, dx$$

where this converges. Under suitable conditions, the theorem states that

$$B(x) = \frac{1}{2\pi i} \int_{L_c} x^s g(s)\, ds. \tag{3.16}$$

(In the usual terminology, the Mellin transform of $B(x)$ is $g(-s)$ rather than $g(s)$.) If $s = \sigma + i\lambda$, the substitution $x = e^t$ gives

$$g(s) = \int_{-\infty}^{\infty} e^{-i\lambda t} C(t)\, dt,$$

where $C(t) = e^{-\sigma t} B(e^t)$. This is the Fourier transform of $C(t)$, and in this way the Mellin inversion theorem can be equated with the Fourier inversion theorem, which may be more familiar to some readers.

Clearly, (3.15) is a case of (3.16), and with some work one can see that the other identities are also cases of it, with $B(x)$ taken to be the function on the right-hand side each time. This certainly places these statements in the wider mathematical landscape, but, for the result we actually want, theorem 3.2.3, it is simpler and more direct to proceed as we did.

Exercises

1 Write down the value of $\displaystyle \frac{1}{2\pi i} \int_{L_c} \frac{x^s}{s^{k+1}}\, ds$, where $c > 0$ and $k \geq 1$ is an integer.

2 Modify the proof of 3.2.3 to obtain formula (3.13). (You need not bother to justify termwise integration.)

3 By considering $f(s) - \alpha\zeta(s)$, show that theorem 3.2.5 can be deduced from the special case in which $\alpha = 0$ (removing the need for 3.2.1).

4 Let $x > 0$, $x \neq 1$ and $c > 0$. By integrating on a rectangle with vertices $b \pm iT$, $c \pm iT$, and letting b tend to $-\infty$ or ∞, prove that

$$\frac{1}{2\pi i} \int_{c-iT}^{c+iT} \frac{x^s}{s}\, ds = E(x) + r_0(x, T),$$

where $r_0(x,T)| \leq x^c/(\pi T|\log x|)$. Deduce that if $\sum_{n=1}^{\infty} a(n)/n^s = f(s)$ is absolutely convergent for $s = c$, then for non-integer $x > 1$,

$$\frac{1}{2\pi i} \lim_{T\to\infty} \int_{c-iT}^{c+iT} \frac{x^s}{s} f(s)\, ds = A(x).$$

3.3 An alternative method: Newman's proof

In this section we present an interesting alternative proof of a version of theorem 3.2.5, in which (FT3) is replaced by a different condition. The method was devised by Newman in 1980 and further simplified by Korevaar in 1982. It actually delivers a prior result from which our 3.2.5 is an easy deduction. In fact, this prior result was already proved by Ingham in 1935, but by a much more complicated method. It is sometimes called the *Ingham-Newman Tauberian theorem*: it is "Tauberian" in the sense that it provides information about the integral of a function $B(x)$, given information about its Mellin (or Laplace) transform.

Like our first method, the Newman-Korevaar proof is still based on complex integration and the Riemann-Lebesgue lemma (so the reader will need this result from section 3.2), but the inversion process is bypassed in favour of an ingeniously chosen integral.

As before, we shall not try to formulate the weakest possible integrability conditions. We simply assume the following condition, which is clearly satisfied by summation functions:

(int) *$B(x)$ is continuous except at integers, and has left and right limits at each integer.*

We now state the Ingham-Newman theorem. Given that it implies our fundamental theorems, the statement seems remarkably innocuous: a function $g(s)$ is defined by a certain formula for $\text{Re } s > 0$ and is assumed holomorphic at 0: the conclusion is simply that the formula is valid at $s = 0$. The real message is that Tauberian theorems are far from trivial! The term "region" will now be used to mean "open set".

Theorem 3.3.1 *Suppose that $B(x)$ is a function (real or complex-valued) that satisfies condition (int), and that, for some M, we have $|B(x)| \leq M/x$ on $[1, \infty)$. Suppose that*

$$g(s) = \int_1^\infty \frac{B(x)}{x^s}\, dx$$

for $\text{Re } s > 0$, and that $g(s)$ can be extended to a function (still denoted by $g(s)$) holomorphic on a region E including $\{s : \text{Re } s \geq 0\}$. Then

$$\int_1^\infty B(x)\, dx = g(0).$$

Proof We show first that it is sufficient to prove the case when $g(0) = 0$.

Suppose this done, and that $B(x)$ and $g(s)$ are given. Define $B_0(x) = B(x) - c/x^2$, where $c = g(0)$. Then, for Re $s > 0$,

$$\int_1^\infty \frac{B_0(x)}{x^s}\,dx = g(s) - \frac{c}{s+1}.$$

Denote this by $g_0(s)$. Then $g_0(0) = g(0) - c = 0$, so we can conclude that

$$\int_1^\infty B_0(x)\,dx = 0.$$

The required statement follows, since

$$\int_1^\infty B_0(x)\,dx = \int_1^\infty B(x)\,dx - c\int_1^\infty \frac{1}{x^2}\,dx = \int_1^\infty B(x)\,dx - c.$$

Assume, then, that $g(0) = 0$, so that $g(s)/s$ is holomorphic at 0. Fix $X > 1$, and let

$$g_X(s) = \int_1^X x^{-s} B(x)\,dx.$$

Then g_X is holomorphic on the whole complex plane (for a proof, see appendix D) and $g_X(0) = \int_1^X B(x)\,dx$. We have to show that $g_X(0) \to 0$ as $X \to \infty$.

Let $\varepsilon > 0$ be given. Choose R such that $M/R \leq \varepsilon$. Let C be the circle $C(0, R)$. Now comes the inventive step. Consider the function

$$J(s) = \frac{1}{s}\left(1 + \frac{s^2}{R^2}\right) X^s.$$

Recall that $X^s = e^{as}$, where $a = \log X$. Hence $J(s)$ is of the form

$$\left(\frac{1}{s} + \frac{s}{R^2}\right)(1 + as + \tfrac{1}{2}a^2s^2 + \cdots) = \frac{1}{s} + K(s),$$

where $K(s)$ is holomorphic everywhere. (In other words, $J(s)$ has a simple pole at 0, with residue 1). Hence by Cauchy's integral theorem and formula,

$$\frac{1}{2\pi i}\int_C J(s) g_X(s)\,ds = g_X(0).$$

Now let C^+ and C^-, respectively, be the portions of C with Re $s \geq 0$ and Re $s \leq 0$. Also, let L be the line segment from $-iR$ to iR. Now $J(s)g(s)$ is holomorphic on E (since $g(s)/s$ is holomorphic at 0). By Cauchy's thorem, it follows that

$$\int_L J(s)g(s)\,ds - \int_{C^+} J(s)g(s)\,ds = 0.$$

By adding this zero quantity to the previous expression for $g_X(0)$, we obtain

$$g_X(0) = I_1(X) + I_2(X) + I_3(X),$$

where

$$I_1(X) = \frac{1}{2\pi i}\int_{C+} J(s)[g_X(s) - g(s)]\,ds,$$

$$I_2(X) = \frac{1}{2\pi i}\int_{C-} J(s)g_X(s)\,ds,$$

$$I_3(X) = \frac{1}{2\pi i}\int_{L} J(s)g(s)\,ds.$$

We now estimate these three integrals separately. The subtlety is that the X^σ occurring in $|J(s)|$ is cancelled by $g_X(s) - g(s)$ on C^+ and by $g_X(s)$ on C^-. Note first that if $s = \sigma + it$ and $|s| = R$ (so that $s\bar{s} = R^2$), then

$$\frac{1}{s} + \frac{s}{R^2} = \frac{1}{R^2}(\bar{s} + s) = \frac{2\sigma}{R^2}.$$

If $\sigma > 0$, then $|x^{-s}B(x)| \le M/x^{\sigma+1}$, so

$$|g(s) - g_X(s)| \le \int_X^\infty \frac{M}{x^{\sigma+1}}\,dx = \frac{M}{\sigma X^\sigma}.$$

So when $|s| = R$ and $\sigma > 0$, we have

$$|J(s)[g(s) - g_X(s)]| \le \frac{2\sigma X^\sigma}{R^2}\,\frac{M}{\sigma X^\sigma} = \frac{2M}{R^2}.$$

By continuity, this inequality also holds at the points $\pm iR$. Hence

$$|I_1(X)| \le \frac{\pi R}{2\pi}\,\frac{2M}{R^2} = \frac{M}{R}.$$

If $\sigma < 0$, then

$$|g_X(s)| \le M\int_1^X x^{-\sigma-1}\,dx < \frac{MX^{-\sigma}}{|\sigma|},$$

so, when $|s| = R$ and $\sigma < 0$, we have

$$|J(s)g_X(s)| \le \frac{2|\sigma|X^\sigma}{R^2}\,\frac{MX^{-\sigma}}{|\sigma|} = \frac{2M}{R^2}.$$

Hence, as before,

$$|I_2(X)| \le \frac{M}{R}.$$

Finally,

$$I_3(X) = \frac{1}{2\pi} \int_{-R}^{R} \frac{g(it)}{it} \left(1 - \frac{t^2}{R^2}\right) X^{it}\, dt.$$

Since $g(it)/it$ is continuous at 0, the Riemann-Lebesgue lemma 3.2.4 (for a bounded interval) shows that $I_3(X) \to 0$ as $X \to \infty$.

So for large enough X, we have $|g_X(0)| \leq 3\varepsilon$, as required. □

Corollary 3.3.2 *Under the same conditions, if*

$$g_1(s) = \int_1^\infty \frac{B(x)}{x^{s-1}}\, dx$$

for Re $s > 1$, *and if* $g_1(s)$ *can be extended to be holomorphic on a region including* $\{s : \text{Re } s \geq 1\}$, *then* $\int_1^\infty B(x)\, dx = g_1(1)$.

Proof This is simply 3.3.1 applied to the function $g(s) = g_1(s+1)$. □

Before giving our new variant of 3.2.5, we give the corresponding statement for Dirichlet *integrals* rather than series, which takes a rather simpler form.

Proposition 3.3.3 *Suppose that f is a complex function differentiable on a region including* $\{s : \text{Re } s \geq 1\}$, *except possibly at the point 1, and that:*

(i) $f(s) = \int_1^\infty \frac{A(x)}{x^{s+1}}\, dx$ *for* Re $s > 1$;

(ii) $f(s) = \frac{\alpha}{s-1} + g(s)$, *where g is differentiable at 1;*

(iii) *there exists* M *such that* $|A(x)| \leq Mx$ *for all* $x \geq 1$.

Then $\int_1^\infty \frac{A(x) - \alpha x}{x^2}\, dx$ *converges to* $g(1)$.

Proof Let

$$B(x) = \frac{A(x)}{x^2} - \frac{\alpha}{x}.$$

Then $|B(x)| \leq (M + |\alpha|)/x$ for $x \geq 1$. For Re $s > 1$, we have

$$\begin{aligned} \int_1^\infty \frac{B(x)}{x^{s-1}}\, dx &= \int_1^\infty \left(\frac{A(x)}{x^{s+1}} - \frac{\alpha}{x^s}\right) dx \\ &= f(s) - \frac{\alpha}{s-1} \\ &= g(s). \end{aligned}$$

The statement follows, by 3.3.2. □

Finally, we derive our variant of 3.2.5.

Theorem 3.3.4 *Suppose that f is a complex function differentiable on a region including $\{s : \mathrm{Re}\, s \geq 1\}$, except possibly at the point 1, and that:*

(FT1) *the series* $\displaystyle\sum_{n=1}^{\infty} \frac{a(n)}{n^s}$ *converges to $f(s)$ when $\mathrm{Re}\, s > 1$;*

(FT2) $f(s) = \dfrac{\alpha}{s-1} + \alpha_0 + (s-1)h(s)$, *where h is differentiable at 1;*

(FT3′) *there exists M such that $|A(x)| \leq Mx$ for all $x \geq 1$.*

Then $\displaystyle\int_1^\infty \frac{A(x) - \alpha x}{x^2}\, dx$ *converges to $\alpha_0 - \alpha$.*

Proof When $\mathrm{Re}\, s > 1$, we have $x^{-s}A(x) \to 0$ as $x \to \infty$, so by 1.3.8, $f(s) = sf_1(s)$, where

$$f_1(s) = \int_1^\infty \frac{A(x)}{x^{s+1}}\, dx.$$

(Note that condition (FT3′) is used again here.) Then

$$\begin{aligned} f_1(s) = \frac{f(s)}{s} &= \alpha\left(\frac{1}{s-1} - \frac{1}{s}\right) + \frac{\alpha_0}{s} + (s-1)\frac{h(s)}{s} \\ &= \frac{\alpha}{s-1} + g(s), \end{aligned}$$

where g is differentiable at 1 and $g(1) = \alpha_0 - \alpha$. The statement follows, by applying 3.3.3 to $f_1(s)$. □

The difference between 3.3.4 and 3.2.5 is that (FT3), a condition on $f(s)$, has been replaced by (FT3′), a condition on $A(x)$. In this sense, 3.3.4 is less purely a result deducing properties of $A(x)$ from those of $f(s)$. If $|a(n)| \leq 1$ for all n, then (FT3′) is trivial, while (FT3) may require some work (as in the case of $\mu(n)$ and $\zeta(s)$). Conversely, there are cases where (FT3) follows from our earlier results, while (FT3′) is not easy to verify, for example,

$$\frac{\zeta'(s)}{\zeta(s)^2} = \sum_{n=1}^{\infty} \frac{\mu(n) \log n}{n^s}.$$

For the purpose of error estimates, one needs the expression found in the proof of 3.2.5 for the integral of $A(x)/x^2 - \alpha/x$ on bounded intervals. Exercise 2 shows how to recover this expression by Newman's method.

Further notes (1) We have presented the Ingham-Newman theorem in terms of the Mellin transform $g(s) = \int_1^\infty x^{-s} B(x)\, dx$. Such theorems are often stated in terms of the Laplace transform

$$G(s) = \int_0^\infty e^{-st} C(t)\, dt$$

(assumed valid for Re $s > 0$). The function $C(t)$ is assumed to be bounded, and the conclusion is that

$$\int_0^\infty C(t)\, dt = G(0).$$

(Condition (int) must be modified to allow the discontinuities to be at points other than integers.) Either version is readily obtained from the other by substitution (see exercise 1). Also, routine modifications to the proof of 3.3.1 will give a direct proof of the Laplace version.

(2) Historically, the approach to the prime number theorem via Tauberian theorems became established in the 1930s. The Ingham-Newman theorem is, roughly, a weak version of the more general "Wiener-Ikehara" Tauberian theorem, known proofs of which are quite hard.

(3) In the accounts of Newman's method known to the author, the integral statement of 3.3.4 is not presented as a theorem in its own right, with the result that there is no mention of the derivation of the series version of our fundamental theorem.

Exercises

1. Derive the Laplace version of the Ingham-Newman theorem from the Mellin version by substituting $x = e^t$ and choosing the function $B(x)$ suitably.

2. *(The expression needed for error estimates)* In 3.3.1, suppose that $g(0) = 0$ and (in addition to the other conditions) $|g(\pm it)| \leq Ct^{-r}$ for all $t \geq 1$, where $r > 0$. By letting R tend to infinity with X fixed, show that

$$\int_1^X B(x)\, dx = \frac{1}{2\pi} \int_{-\infty}^\infty \frac{X^{it}}{it} g(it)\, dt.$$

Formulate the corresponding variant of 3.3.2 (but not 3.3.3). In 3.3.4, assume the additional condition $|f(1 \pm it)| \leq Ct^{1-r}$ for $t \geq 1$. By taking $B(x)$ to be $[A(x) - \alpha x - (\alpha_0 - \alpha)]/x^2$, derive the identity

$$\int_1^X \frac{A(x) - \alpha x}{x^2}\, dx = (\alpha_0 - \alpha)\left(1 - \frac{1}{X}\right) + \frac{1}{2\pi i}\int_{L_1} X^{s-1}\frac{h(s)}{s}\, ds$$

(as found, under different conditions, in the proof of 3.2.5).

3.4 The limit and series versions of the fundamental theorem; the prime number theorem

The limit and series versions

We have already obtained the *integral* version of the fundamental theorem for Dirichlet series, in the alternative forms 3.2.5 and 3.3.4. We will now deduce the *limit* version, stating (in the same notation) that

$$\frac{A(x)}{x} \to \alpha \quad \text{as } x \to \infty$$

and at the same time the *series* version, stating that

$$\sum_{n=1}^{\infty} \frac{a(n) - \alpha}{n}$$

converges (to $\alpha_0 - \gamma\alpha$). We can then read off what these statements say for special choices of Dirichlet series. In particular, when $a(n) = \Lambda(n)$, the limit version will give the desired fact that $\lim_{x\to\infty} \psi(x)/x = 1$. So the prime number theorem is obtained as just one example of a family of statements of this kind (though certainly the most important one).

The connecting link is provided by the following result, a Tauberian theorem of a much more elementary type than the Ingham-Newman theorem.

Proposition 3.4.1 *Suppose that A is a real-valued function on $[1, \infty)$ such that, for some α,*

$$\int_1^{\infty} \frac{A(x) - \alpha x}{x^2}\, dx$$

is convergent. Suppose also that

either (i) *A is increasing and non-negative,*

or (ii) *there is another function B such that B and $B - A$ are increasing and non-negative and for some β, $\int_1^{\infty} [B(x) - \beta x]/x^2\, dx$ converges.*

Then $\dfrac{A(x)}{x} \to \alpha$ *as* $x \to \infty$.

Proof Statement (ii) follows at once from (i) because we then have

$$\frac{B(x)}{x} \to \beta \quad \text{and} \quad \frac{B(x) - A(x)}{x} \to \beta - \alpha \quad \text{as } x \to \infty,$$

and hence $A(x)/x \to \alpha$.

We prove (i) first for the case $\alpha \neq 0$. It is enough to prove it when $\alpha = 1$ (for other cases, consider $A(x)/\alpha$). Let $0 < \delta < \frac{1}{2}$. Convergence of the stated integral implies that, for any $\delta_1 > 0$, there exists R such that

$$\left| \int_{x_0}^{x_1} \frac{A(x) - x}{x^2} \, dx \right| < \delta_1$$

whenever $x_1 > x_0 \geq R$. Suppose that for some $x_0 > R$ we have $A(x_0) > (1+\delta)x_0$. Since $A(x)$ is increasing, $A(x) > (1+\delta)x_0$ for all $x \geq x_0$. Let $x_1 = (1+\delta)x_0$. Then

$$\begin{aligned} \int_{x_0}^{x_1} \frac{A(x) - x}{x^2} \, dx &> (1+\delta)x_0 \int_{x_0}^{x_1} \frac{1}{x^2} \, dx - \int_{x_0}^{x_1} \frac{1}{x} \, dx \\ &= x_1 \left(\frac{1}{x_0} - \frac{1}{x_1} \right) - \log \frac{x_1}{x_0} \\ &= \delta - \log(1+\delta). \end{aligned}$$

Now take $\delta_1 = \delta - \log(1+\delta)$. Then $\delta_1 > 0$ (in fact, from the series for $\log(1+\delta)$, one has $\delta_1 \geq \frac{1}{3}\delta^2$). So if R corresponds to δ_1 as above, then $A(x) \leq (1+\delta)x$ for all $x > R$.

Similarly, if $A(x_0) < (1-\delta)x_0$ for some $x_0 \geq 2R$, we take $x_2 = (1-\delta)x_0$ (note that $x_2 \geq R$), and we find that

$$\int_{x_2}^{x_0} \frac{A(x) - x}{x^2} \, dx \leq -\delta_2,$$

where $\delta_2 = -\log(1-\delta) - \delta$ ($> \frac{1}{2}\delta^2$). If R corresponds to δ_2 as before, we deduce that $A(x) \geq (1-\delta)x$ for all $x \geq 2R$.

Finally, the case $\alpha = 0$ is easy. Given δ, there exists R such that

$$\int_{x_0}^{\infty} \frac{A(x)}{x^2} \, dx < \delta$$

whenever $x_0 \geq R$. Suppose that $A(x_0) \geq \delta x_0$ for some x_0. Then

$$\int_{x_0}^{\infty} \frac{A(x)}{x^2} \, dx \geq \delta x_0 \int_{x_0}^{\infty} \frac{1}{x^2} \, dx = \delta.$$

So we must have $A(x) \leq \delta x$ for $x \geq R$. □

We can now derive the limit and series statements, as promised. Actually, the series statement itself can be expressed in either of two equivalent forms, both of which have their uses. Conditions (FT1), (FT2), (FT3) and (FT3′) are as stated in theorems 3.2.5 and 3.3.4.

Theorem 3.4.2 *Suppose that $a(n)$ (real or complex) and $f(s)$ satisfy conditions (FT1), (FT2) and (FT3), and that*

either (FT4) $a(n) \geq 0$ *for all* n,

or (FT5) *there exist* $b(n)$, $g(s)$ *satisfying conditions (FT1), (FT2) and (FT3), with* $|a(n)| \leq b(n)$ *for all* n.

Then:

(i) $\dfrac{A(x)}{x} \to \alpha$ *as* $x \to \infty$,

(ii) $\displaystyle\sum_{n \leq x} \frac{a(n)}{n} - \alpha \log x \to \alpha_0$ *as* $x \to \infty$.

(iii) $\displaystyle\sum_{n=1}^{\infty} \frac{a(n) - \alpha}{n} = \alpha_0 - \gamma\alpha$, *where* γ *is Euler's constant.*

Condition (FT3) can be replaced by (FT3′) each time.

Proof By 3.2.5 and 3.3.4, the integral hypothesis of 3.4.1 is satisfied. Under condition (FT4), $A(x)$ is increasing, so $\lim_{x\to\infty}[A(x)/x] = \alpha$. Under condition (FT5), if $a(n)$ is real, then $B(x)$ and $B(x) - A(x)$ are increasing, so the second version of 3.4.1 applies to give the same conclusion. If $a(n)$ is complex, we obtain the same result by applying this to Re $A(x)$ and Im $A(x)$ separately.

By Abel's summation formula (1.3.6), we now have in either case

$$\begin{aligned}\sum_{n \leq x} \frac{a(n)}{n} - \alpha \log x &= \frac{A(x)}{x} + \int_1^x \frac{A(y)}{y^2}\,dy - \int_1^x \frac{\alpha}{y}\,dy \\ &= \frac{A(x)}{x} + \int_1^x \frac{A(y) - \alpha y}{y^2}\,dy \\ &\to \alpha + (\alpha_0 - \alpha) = \alpha_0 \quad \text{as } x \to \infty.\end{aligned}$$

Hence also

$$\begin{aligned}\sum_{n \leq x} \frac{a(n) - \alpha}{n} &= \sum_{n \leq x} \frac{a(n)}{n} - \alpha \log x - \alpha \left(\sum_{n \leq x} \frac{1}{n} - \log x \right) \\ &\to \alpha_0 - \gamma\alpha \quad \text{as } x \to \infty.\end{aligned}$$

□

Remark 1 For the purposes of the limit statement (i), there is no need to know α_0 explicitly.

Remark 2 In the case when $f(s)$ has no pole at 1 (so that $\alpha = 0$), α_0 is simply $f(1)$, and the series statement (in either form) becomes $\sum_{n=1}^{\infty} a(n)/n = f(1)$. In other words, the original Dirichlet series is also valid at $s = 1$.

Remark 3 In the special cases considered below, we will check both the alternative conditions (FT3) and (FT3′), so that either form of the integral theorem is sufficient for the conclusion.

The prime number theorem

We have now reached our original objective.

Theorem 3.4.3 (the prime number theorem) *Let ψ be Chebyshev's function. Then*

$$\frac{\psi(x)}{x} \to 1 \quad \textit{as } x \to \infty,$$

$$\pi(x) \sim \mathrm{li}(x) \quad \textit{as } x \to \infty.$$

Proof We show that the first statement is a case of 3.4.2. As we saw in 1.6.2 and 2.4.5, the second statement then follows. Let $f(s) = -\zeta'(s)/\zeta(s)$. Then

$$f(s) = \sum_{n=1}^{\infty} \frac{\Lambda(n)}{n^s} \quad \text{for Re } s > 1$$

and

$$f(s) = \frac{1}{s-1} - \gamma + (s-1)h(s),$$

with $h(s)$ differentiable at 1. Of course, $\Lambda(n) \geq 0$ for all n, and $\sum_{n \leq x} \Lambda(n) = \psi(x)$. To verify (FT3), note that, by 3.1.8 and 3.1.14, $|f(\sigma \pm it)| \leq P(t)$ when $\sigma \geq 1$, $t \geq 2$, where $P(t) = 2(\log t + 5)^9$. The integral $\int_1^{\infty} P(t)/t^2 \, dt$ is convergent, as required. Alternatively, Chebyshev's upper estimate $\psi(x) \leq 2x$ (2.4.6) shows that (FT3′) is satisfied. □

The reader may be tempted to pause for celebration!

As we saw in section 1.5, it follows that equally

$$\pi(x) \sim \frac{x}{\log x} \qquad \text{and} \qquad \pi(x) \sim \frac{x}{\log x - 1}.$$

The prime number theorem is often quoted in one of these forms (especially the first one).

An obvious piece of unfinished business is the *rate* of convergence. We return to this question in chapter 5, where we will show that $|\pi(x) - \mathrm{li}(x)|$ is ultimately small compared with $x/(\log x)^k$ for every positive k. Once this is known, it is clear that $\mathrm{li}(x)$ is a better approximation than $x/(\log x - 1)$, which in turn is better than $x/\log x$.

Needless to say, the prime number theorem has interesting consequences and applications. We describe some in section 3.5.

At the same time, we have of course obtained integral and series statements relating to the Dirichlet series $\sum_{n=1}^{\infty} \Lambda(n)/n^s$, as a particular case of the general theorems. We record these statements here.

Proposition 3.4.4 *For the Chebyshev function ψ and the von Mangoldt function $\Lambda(n)$, we have:*

$$\int_1^{\infty} \frac{\psi(x) - x}{x^2}\, dx = -\gamma - 1,$$

$$\sum_{n \le x} \frac{\Lambda(n)}{n} - \log x \to -\gamma \quad \textit{as } x \to \infty.$$

$$\sum_{n=1}^{\infty} \frac{\Lambda(n) - 1}{n} = -2\gamma.$$

□

Further special cases of the basic theorems

The second most important case of the basic theorems 3.2.5 and 3.4.2 is found by applying them to $1/\zeta(s)$. This time we need condition (FT5) instead of (FT4).

Theorem 3.4.5 *Let $\mu(n)$ be the Möbius function and $M(x) = \sum_{n \le x} \mu(n)$. Then*

$$\int_1^{\infty} \frac{M(x)}{x^2}\, dx = 0,$$

$$\frac{M(x)}{x} \to 0 \quad \textit{as } x \to \infty,$$

$$\sum_{n=1}^{\infty} \frac{\mu(n)}{n} = 0.$$

Proof Let $f(s) = 1/\zeta(s)$ for $s \neq 1$, and $f(1) = 0$. Then

$$f(s) = \sum_{n=1}^{\infty} \frac{\mu(n)}{n^s} \quad \text{for Re } s > 1$$

and $f(s) = 0 + (s-1)h(s)$, with $h(s)$ differentiable at 1. By 3.1.14, condition (FT3) is satisfied, with $P(t) = 4(\log t + 5)^7$. The alternative condition (FT3′) is trivial, since obviously $|M(x)| \leq x$. Also, since $|\mu(n)| \leq 1$ for all n, we can take $b(n) = 1$ and $g(s) = \zeta(s)$ in (FT5): these functions clearly satisfy the required conditions. The statements follow. □

Similar statements apply to the Liouville function:

Proposition 3.4.6 *Let $\lambda(n)$ be the Liouville function, and let $S_\lambda(x) = \sum_{n \leq x} \lambda(n)$. Then*

$$\int_1^{\infty} \frac{S_\lambda(x)}{x^2}\, dx = \lim_{x \to \infty} \frac{S_\lambda(x)}{x} = \sum_{n=1}^{\infty} \frac{\lambda(n)}{n} = 0.$$

Proof Let $f(s) = \zeta(2s)/\zeta(s)$ for $s \neq 1$ and $f(1) = 0$. Then

$$f(s) = \sum_{n=1}^{\infty} \frac{\lambda(n)}{n^s} \quad \text{for Re } s > 1$$

and f is holomorphic, with value 0, at 1. For condition (FT3), note that $|\zeta(2s)| \leq \zeta(2)$ when $\sigma \geq 1$. Again, (FT3′) is immediate. The statements follow, as before. □

Because $1/\zeta(s)$ has no pole at 1, we can extend these results to the whole line Re $s = 1$, as follows. We confine ourselves to the Möbius function (the Liouville function is similar), and to stating the series version, which amounts to saying that the original Dirichlet series converges on this line.

Proposition 3.4.7 *For all non-zero, real t, we have*

$$\sum_{n=1}^{\infty} \frac{\mu(n)}{n^{1+it}} = \frac{1}{\zeta(1+it)}.$$

Proof Fix t. Let $a(n) = n^{-it}\mu(n)$ and $f(s) = 1/\zeta(s+it)$. Then $f(s)$ is defined and differentiable on the whole line Re $s = 1$, and $\sum_{n=1}^{\infty}[a(n)/n^s] = f(s)$ for Re $s > 1$. As for any holomorphic function, we have $f(s) = f(1) + (s-1)h(s)$, where $h(s)$ is differentiable at 1, so (in the notation of 3.4.2) $\alpha = 0$ and $\alpha_0 = f(1)$. Also, if $|u| \geq |t| + 2$, then

$$|f(\sigma + iu)| \leq 4[\log(|t| + |u| + 5)]^7,$$

so condition (FT3) is satisfied; again, (FT3′) is trivial. As before, we can take $b(n) = 1$ in condition (FT5). The statement follows. □

In the same way, we can extend 3.4.4 by working with the function

$$f(s) = \zeta(s) + \frac{\zeta'(s)}{\zeta(s)}.$$

By adding the two series, one sees that this function has no pole at 1. With t fixed, take

$$f_t(s) = f(s + it) = \sum_{n=1}^{\infty} \frac{a(n)}{n^s},$$

where $a(n) = n^{-it}[1 - \Lambda(n)]$. The dominating sequence is $b(n) = 1 + \Lambda(n)$. We obtain:

Proposition 3.4.8 *Let $f(s)$ be as just stated. For all non-zero, real t, we have*

$$\sum_{n=1}^{\infty} \frac{1 - \Lambda(n)}{n^{1+it}} = f(1 + it). \qquad \square$$

Corollary 3.4.9 *For $t \neq 0$, we have*

$$\sum_{n=1}^{N} \frac{\Lambda(n)}{n^{1+it}} = -\frac{\zeta'(1+it)}{\zeta(1+it)} - \frac{1}{it}N^{-it} + r_N,$$

where $r_N \to 0$ as $N \to \infty$.

Proof Recall that, by 3.1.3,

$$\sum_{n=1}^{N} \frac{1}{n^{1+it}} = \zeta(1 + it) - \frac{1}{it}N^{-it} + q_N,$$

where $q_N \to 0$ as $N \to \infty$. The result follows by adding the sum of N terms in 3.4.8. □

Hence the partial sums of $\sum \Lambda(n)/n^{1+it}$ eventually describe circles around the point $-(\zeta'/\zeta)(1 + it)$ in the same way that those of $\sum 1/n^{1+it}$ do so around $\zeta(1 + it)$.

In particular, these partial sums are bounded. Recall that, for $\operatorname{Re} s > 1$, we have $\log \zeta(s) = \sum_{n=1}^{\infty} c(n)/n^s$, where $c(n) = \Lambda(n)/\log n$. Since

$$\frac{c(n)}{n^s} = \frac{\Lambda(n)}{n^s} \frac{1}{\log n},$$

discrete Abel summation (in the form 1.3.4) shows that $\sum_{n=1}^{\infty} c(n)/n^s$ converges for $s \neq 1$ with $\mathrm{Re}\, s = 1$. A routine argument shows that the sum of the series is continuous onto this line, from which it follows that it still equals $\log \zeta(s)$ there. It follows easily that the Euler product converges to $\zeta(s)$ for such s. (For the details, see, e.g., [Titch], section 3.14.)

Exercises

1 Prove that $\psi(x) - \theta(x) \sim x^{1/2}$ as $x \to \infty$.

2 Given that $\sum_{n=1}^{\infty} [a(n) - \alpha]/n$ converges, use an exercise from section 1.3 to show that

$$\frac{A(x)}{x} \to \alpha \quad \text{as } x \to \infty.$$

Deduce that statement (ii) of theorem 3.4.2 implies both the limit and integral forms of the fundamental theorem.

3 Let $M_2(x) = \sum_{n \leq x} |\mu(n)|$. Use the theorems of the present section to re-prove 2.5.5 in the form

$$\frac{M_2(x)}{x} \to \frac{1}{\zeta(2)} = \frac{6}{\pi^2} \quad \text{as } x \to \infty.$$

4 Let $a(n) = \mu(n) \log n$. What is the function that has Dirichlet series $\sum_{n=1}^{\infty} a(n)/n^s$? Show that

$$\int_1^{\infty} \frac{A(x)}{x^2}\, dx = -1.$$

5 Prove that $\sum_{p \in P} 1/p^s$ is convergent for $s \neq 1$ with $\mathrm{Re}\, s = 1$.

6 Let E_k be the set of integers n such that $2^{k-1} < n \leq 2^k$. Define $a(n)$ to be 1 if n is in $\bigcup_{k=1}^{\infty} E_{2k}$ and 0 otherwise. Let $f(s) = \sum_{n=1}^{\infty} a(n)/n^s$ for $\mathrm{Re}\, s > 1$. By applying Euler's summation formula to $\sum_{n \in E_{2k}} 1/n^s$, show that $f(s)$ can be expressed as $g(s) + h(s)$, where

$$g(s) = \frac{1}{(s-1)(2^{s-1}+1)}$$

and $h(s)$ is holomorphic for $\mathrm{Re}\, s > 0$ (so that $f(s)$ has a simple pole at 1). Show however that $A(N)/N$ does not tend to a limit as $N \to \infty$, by considering $N = 2^{2k}$ and $N = 2^{2k+1}$. Which of the hypotheses of the fundamental theorems is not satisfied?

7 Show by the following steps that the prime number theorem (if proved another way) implies that $\zeta(1+it) \neq 0$ for all $t \neq 0$. Suppose that $\zeta(s)$ has a zero of order m at $s_0 = 1 + it_0$. Let

$$f(s) = s \int_1^\infty \frac{\psi(x) - x}{x^{s+1}} \, dx.$$

Show that $(s - s_0)f(s) \to m$ as $s \to s_0$. However, assuming that $\psi(x) \sim x$, show that $(\sigma - 1)|f(\sigma + it_0)|$ tends to 0 as $\sigma \to 1^+$, hence obtaining a contradiction.

3.5 Some applications of the prime number theorem

Primes in intervals

Having worked so hard to establish the prime number theorem, one would hope to see some interesting applications. We start with an easy application to the number of primes in an interval $(x, cx]$, where $c > 1$. Of course, this number is $\pi(cx) - \pi(x)$.

First, some remarks about the symbol $\sim$. We consider only strictly positive functions. Recall that $f(x) \sim g(x)$ as $x \to \infty$ means that $f(x)/g(x) \to 1$. The following facts follow easily (in each case, $x \to \infty$):

if $f(x) \sim g(x)$ *and* $g(x) \sim h(x)$, *then* $f(x) \sim h(x)$;

if $f(x) \sim ah(x)$ *and* $g(x) \sim bh(x)$, *where* $a > b > 0$, *then* $f(x) \pm g(x) \sim (a \pm b)h(x)$;

if $f(x) \sim f_0(x)$ *and* $g(x) \sim g_0(x)$, *then* $\dfrac{f(x)}{g(x)} \sim \dfrac{f_0(x)}{g_0(x)}$.

Using the prime number theorem in its least accurate form, $\pi(x) \sim (x/\log x)$, we deduce:

Proposition 3.5.1 *Let* $c > 1$. *When* $x \to \infty$, *we have*

$$\pi(cx) \sim c\pi(x),$$

$$\pi(cx) - \pi(x) \sim (c-1)\frac{x}{\log x}.$$

Proof Clearly,

$$\log(cx) = \log x + \log c \sim \log x \quad \text{as } x \to \infty.$$

Hence

$$\pi(cx) \sim \frac{cx}{\log cx} \sim c\frac{x}{\log x}.$$

Both statements follow, by the principles just listed. □

Corollary 3.5.2 *Let $c > 1$. Then, for all large enough x, there are primes in the interval $(x, cx]$.*

Proof By 3.5.1,

$$(\pi(cx) - \pi(x))\frac{\log x}{x}$$

tends to the limit $c - 1$ as $x \to \infty$; hence it is certainly non-zero for all large enough x. □

In the sense of *ratios*, 3.5.1 (with $c = 2$) says that there are about as many primes in $[x, 2x]$ as there are in $[1, x]$. This may seem to contradict our (quite correct) impression that the primes thin out as x increases. There is no contradiction, since we are talking about the ratio of two numbers that themselves tend to infinity. The primes thin out very slowly in the same way that the function $\log x$ grows very slowly: it increases, but $(\log 2x)/(\log x)$ tends to 1. Once we know the prime number theorem with an error estimate, we will be able to show that the difference, $2\pi(x) - \pi(2x)$, tends to infinity as $x \to \infty$.

As already stated, $\mathrm{li}(x)$ is in fact a better approximation to $\pi(x)$ than $x/(\log x)$, and consequently a better approximation to $\pi(cx) - \pi(x)$ is

$$\int_x^{cx} \frac{1}{\log t}\, dt.$$

Again, the error estimate will make this more precise. For the moment, we just give a numerical table comparing the number of primes in intervals of length 100,000 with the number predicted by $\mathrm{li}(x)$. (The intervals are described in thousands.)

interval	primes	estimate	interval	primes	estimate
0–100	9,592	9,630	500–600	7,560	7,566
100–200	8,392	8,407	600–700	7,445	7,472
200–300	8,013	8,051	700–800	7,408	7,393
300–400	7,863	7,836	800–900	7,323	7,325
400–500	7,678	7,684	900–1000	7,224	7,265

The nth prime number

Proposition 3.5.3 *Let p_n be the nth prime number. Then $p_n \sim n \log n$ as $n \to \infty$.*

Proof Since $\pi(x) \log x/x \to 1$ as $x \to \infty$ and $\pi(p_n) = n$, we have

$$\frac{n \log p_n}{p_n} \to 1 \quad \text{as } n \to \infty. \tag{3.17}$$

Hence

$$\log n + \log \log p_n - \log p_n \to 0 \quad \text{as } n \to \infty. \tag{3.18}$$

Dividing by $\log p_n$, we see that

$$\frac{\log n}{\log p_n} \to 1 \quad \text{as } n \to \infty. \tag{3.19}$$

By multiplying (3.17) and (3.19), we obtain

$$\frac{n \log n}{p_n} \to 1 \qquad \text{as } n \to \infty. \qquad \square$$

Note The better estimate $\pi(x) \sim x/(\log x - 1)$ amends (3.17) to $p_n \sim n(\log p_n - 1)$. By (3.19), $\log \log n - \log \log p_n \to 0$. Using (3.18), we deduce that $\log p_n - (\log n + \log \log n) \to 0$, and hence the corresponding better approximation to p_n is $q(n)$, where

$$q(n) = n(\log n + \log \log n - 1).$$

For a numerical illustration, we use the fact that $\pi(100,000) = 9592$. In fact, if $n = 9592$, then $p_n = 99,991$. We find that $n \log n = 87,949$, while $q(n) = 99,613$.

Enumeration of numbers with k prime factors

Our third application of the prime number theorem is rather more substantial. A natural extension of the original question is to ask, for a given $k \geq 2$, how many integers less than x have k prime factors. We distinguish two version of this question, depending on whether repetitions are allowed. Let $A_k(x)$ be the set of integers $n \leq x$ that are products of k primes (not necessarily distinct), and let $B_k(x)$ be the set of those that are products of k *distinct* primes. In terms of the arithmetic functions ω and Ω:

$$A_k(x) = \{n \leq x : \Omega(n) = k\},$$

$$B_k(x) = \{n \leq x : \Omega(n) = \omega(n) = k\}.$$

Clearly, $B_k(x) \subset A_k(x)$ for $k \geq 2$. Let $\Pi_k(x)$ and $\pi_k(x)$, respectively, be the number of members of $A_k(x)$ and $B_k(x)$. (Warning: Notation varies between different writers.)

We actually prove a k-dimensional extension of the prime number theorem, as follows. Define $E_k(x)$ to be the set of k-tuples $(p_1, \ldots, p_k)$ with each p_j prime and $p_1 p_2 \ldots p_k \leq x$. There is no mystery about repetitions: $(3,2,3)$ and $(2,3,3)$ are different triples. Let $N_k(x)$ be the number of members of $E_k(x)$. Of course, $N_1(x) = \pi(x)$. We will estimate $N_k(x)$ and deduce estimates for $\Pi_k(x)$ and $\pi_k(x)$. To do this, we will require several lemmas.

For each $n \leq x$, let c_n be the number of members $(p_1, \ldots, p_k)$ of $E_k(x)$ (if any) such that $p_1 p_2 \ldots p_k = n$. With x fixed, let us just write A_k for $A_k(x)$, and similarly B_k, E_k. Once all this notation has been digested, the following facts are clear:

$$c_n \leq k! \quad \text{for each } n, \tag{3.20}$$

$$n \in A_k \Leftrightarrow c_n \geq 1, \tag{3.21}$$

$$n \in B_k \Leftrightarrow c_n = k!\,, \tag{3.22}$$

$$N_k(x) = \sum_{n \leq x} c_n = \sum_{n \in A_k} c_n. \tag{3.23}$$

Lemma 3.5.4 *We have* $k!\,\pi_k(x) \leq N_k(x) \leq k!\,\Pi_k(x)$.

Proof By (3.23) and (3.20), $N_k(x) = \sum_{n \in A_k} c_n \leq k!\,\Pi_k(x)$. By (3.22), (3.23) and (3.20),

$$k!\,\pi_k(x) = \sum_{n \in B_k} c_n \leq N_k(x).$$

□

Lemma 3.5.5 *For* $k \geq 2$, *we have* $\Pi_k(x) - \pi_k(x) \leq N_{k-1}(x)$.

Proof Define a mapping ϕ on E_{k-1} by

$$\phi(p_1, p_2, \ldots, p_{k-1}) = p_1^2 p_2 \ldots p_{k-1}.$$

The set of values assumed by ϕ includes all members of $A_k \setminus B_k$ (with many repetitions), as well as some numbers greater than x. □

We now define

$$L_k(x) = \sum \left\{ \frac{1}{p_1 \ldots p_k} : (p_1, \ldots, p_k) \in E_k(x) \right\}$$

and

$$\theta_k(x) = \sum \{\log(p_1 \dots p_k) : (p_1, \dots, p_k) \in E_k(x)\}.$$

Clearly,

$$L_k(x) = \sum_{n \le x} \frac{c_n}{n}, \qquad \theta_k(x) = \sum_{n \le x} c_n \log n.$$

Of course, $\theta_1(x)$ is the usual $\theta(x)$. We shall estimate $\theta_k(x)$ and deduce an estimate for $N_k(x)$ as in the one-dimensional case.

Lemma 3.5.6 *We have* $L_k(x) \sim (\log\log x)^k$ *as* $x \to \infty$.

Proof Note that $L_1(x) = \sum_{p \in P[x]}(1/p)$. If $n = p_1 \dots p_k$ and $n \le x$, then clearly each $p_i \le x$. Conversely, if each $p_i \le x^{1/k}$, then $n \le x$. Hence

$$\left(L_1(x^{1/k})\right)^k \le L_k(x) \le (L_1(x))^k.$$

By 2.6.4, $L_1(x) \sim \log\log x$ as $x \to \infty$. Hence also

$$L_1(x^{1/k}) \sim \log\left(\frac{1}{k}\log x\right) = \log(\log x - \log k) \sim \log\log x \quad \text{as } x \to \infty.$$

The statement follows. □

The basis of the proof is induction on k. The next two lemmas form the essential link with $k-1$. We set $L_0(x) = 1$.

Lemma 3.5.7 *For each* $k \ge 1$, *we have*

$$L_k(x) = \sum_{p \in P[x]} \frac{1}{p} L_{k-1}\left(\frac{x}{p}\right).$$

Proof The case $k = 1$ is the definition. For $k \ge 2$, we have

$$\begin{aligned} L_k(x) &= \sum_{p_k \in P[x]} \frac{1}{p_k} \sum \left\{ \frac{1}{p_1 \dots p_{k-1}} : (p_1, \dots, p_{k-1}) \in E_{k-1}\left(\frac{x}{p_k}\right) \right\} \\ &= \sum_{p_k \in P[x]} \frac{1}{p_k} L_{k-1}\left(\frac{x}{p_k}\right). \end{aligned}$$

□

Lemma 3.5.8 *For each* $k \ge 2$, *we have*

$$(k-1)\theta_k(x) = k \sum_{p \in P[x]} \theta_{k-1}\left(\frac{x}{p}\right).$$

Proof By symmetry,

$$\sum \{\log p_j : (p_1, \ldots, p_k) \in E_k(x)\}$$

has the same value for each j: call it $t_k(x)$. Since $\log(p_1 \ldots p_k) = \log p_1 + \cdots + \log p_k$, we have $\theta_k(x) = kt_k(x)$. In the same way,

$$\sum \{\log(p_1 \ldots p_{k-1}) : (p_1, \ldots, p_k) \in E_k(x)\} = (k-1)t_k(x),$$

so

$$\begin{aligned}(k-1)\theta_k(x) &= k \sum \{\log(p_1 \ldots p_{k-1}) : (p_1, \ldots, p_k) \in E_k(x)\} \\ &= k \sum_{p_k \in P[x]} \sum \left\{\log(p_1 \ldots p_{k-1}) : (p_1, \ldots, p_{k-1}) \in E_{k-1}\left(\frac{x}{p_k}\right)\right\} \\ &= k \sum_{p_k \in P[x]} \theta_{k-1}\left(\frac{x}{p_k}\right).\end{aligned}$$

□

Proposition 3.5.9 *When $x \to \infty$, we have*

$$\theta_k(x) \sim kx(\log\log x)^{k-1}.$$

Proof Let

$$f_k(x) = \theta_k(x) - kxL_{k-1}(x).$$

We shall prove by induction that

$$\frac{f_k(x)}{x(\log\log x)^{k-1}} \to 0 \quad \text{as } x \to \infty.$$

By 3.5.6, it then follows that

$$\theta_k(x) = kxL_{k-1}(x) + f_k(x) \sim kx(\log\log x)^{k-1} \quad \text{as } x \to \infty.$$

Note first that $f_1(x) = \theta(x) - x$, so the case $k = 1$ is the prime number theorem in the form $\theta(x) \sim x$. Assume that the statement is true for a certain $k \geq 1$. By 3.5.7 and 3.5.8,

$$\begin{aligned}kf_{k+1}(x) &= (k+1) \sum_{p \in P[x]} \theta_k\left(\frac{x}{p}\right) - k(k+1)x \sum_{p \in P[x]} \frac{1}{p} L_{k-1}\left(\frac{x}{p}\right) \\ &= (k+1) \sum_{p \in P[x]} f_k\left(\frac{x}{p}\right). \end{aligned} \tag{3.24}$$

Choose $\varepsilon > 0$. By the induction hypothesis, there exists x_0 such that, for all $x \geq x_0$,

$$|f_k(x)| \leq \varepsilon x(\log\log x)^{k-1}. \tag{3.25}$$

Clearly, $|f_k(x)|$ is bounded (say by A) for $2 \leq x \leq x_0$. Also, for large enough x, we have

$$\sum_{p \in P[x]} \frac{1}{p} \leq 2 \log \log x.$$

We now take the sum in (3.24) in two parts. Recall that $P(a, b]$ denotes the set of primes in the interval $(a, b]$. When $p \leq (x/x_0)$, we have $(x/p) \geq x_0$, so (3.25) implies that, for large enough x,

$$\begin{aligned}
\sum_{p \in P[x/x_0]} \left| f_k\left(\frac{x}{p}\right) \right| &\leq \varepsilon (\log \log x)^{k-1} \sum_{p \in P[x/x_0]} \frac{x}{p} \\
&\leq \varepsilon x (\log \log x)^{k-1} \sum_{p \in P[x]} \frac{1}{p} \\
&\leq 2 \varepsilon x (\log \log x)^k.
\end{aligned}$$

When $p > (x/x_0)$, we have $(x/p) < x_0$, so

$$\sum_{p \in P(x/x_0, x]} \left| f_k\left(\frac{x}{p}\right) \right| \leq A \pi(x) \leq A x.$$

For large enough x, we have $A \leq \varepsilon (\log \log x)^k$ and hence

$$\sum_{p \in P[x]} \left| f_k\left(\frac{x}{p}\right) \right| \leq 3 \varepsilon x (\log \log x)^k.$$

With (3.24), this shows that the induction statement holds for $k+1$. □

Proposition 3.5.10 *When $x \to \infty$, we have*

$$N_k(x) \sim \frac{kx}{\log x} (\log \log x)^{k-1}.$$

Proof Let $k \geq 2$ (the case $k = 1$ is the prime number theorem). By Abel summation, together with (3.23), we have

$$\theta_k(x) = \sum_{n \leq x} c_n \log n = N_k(x) \log x - \int_1^x \frac{N_k(t)}{t} \, dt.$$

Now $N_k(t) \leq k!t$, by 3.5.4, so the integral in the above expression is not greater than $k!x$. Dividing by $kx(\log \log x)^{k-1}$, we deduce that

$$\frac{N_k(x) \log x}{kx (\log \log x)^{k-1}} \to 1 \quad \text{as } x \to \infty.$$ □

Note that this argument was actually simpler than the corresponding step for $k = 1$, for which we expressed $\pi(x)$ in terms of $\theta(x)$ (1.6.2). The simplification is made possible by the factor $(\log\log x)^{k-1}$.

Theorem 3.5.11 *For each $k \geq 1$, we have*

$$\Pi_k(x) \sim \pi_k(x) \sim \frac{x}{\log x} \frac{(\log\log x)^{k-1}}{(k-1)!} \quad \text{as } x \to \infty.$$

Proof Again we only need to consider $k \geq 2$. By 3.5.10, $N_{k-1}(x)/N_k(x) \to 0$ as $x \to \infty$. By 3.5.4 and 3.5.5,

$$k!\,\Pi_k(x) - N_k(x) \leq k![\Pi_k(x) - \pi_k(x)] \leq k!\,N_{k-1}(x),$$

so that

$$N_k(x) \leq k!\,\Pi_k(x) \leq N_k(x) + k!\,N_{k-1}(x).$$

On dividing by $N_k(x)$, it is now clear that

$$k!\,\Pi_k(x) \sim N_k(x) \quad \text{as } x \to \infty.$$

The required statement for $\Pi_k(x)$ follows, by 3.5.10. Similar reasoning applies to $\pi_k(x)$. □

The theorem was first proved by Landau in 1900. The simpler proof given here is due to E. M. Wright (1954). It is again instructive to compare some numbers from real life. The following tables give figures for $k = 2, 3$ and 5. We denote by $e_k(x)$ the estimate provided by 3.5.11.

x	$\pi_2(x)$	$\Pi_2(x)$	$e_2(x)$	$\pi_3(x)$	$\Pi_3(x)$	$e_3(x)$
1,000	288	299	280	135	247	270
10,000	2,600	2,625	2,411	1,800	2,569	2,676
100,000	23,313	23,378	21,224	19,919	25,556	25,930
1,000,000	209,867	210,035	190,061	206,964	250,853	249,530

x	$\pi_5(x)$	$\Pi_5(x)$	$e_5(x)$
1,000	0	76	84
10,000	24	963	1,099
100,000	910	11,185	12,901
1,000,000	18,387	124,465	143,372

While the gap between $\Pi_2(x)$ and $\pi_2(x)$ is quite small, it is clear that x has to be very large indeed for there to be any convergence between $\Pi_5(x)$

and $\pi_5(x)$. The remarkably good agreement between $e_3(x)$ and $\Pi_3(x)$ is not typical of other k.

More elaborate expressions are known that give better approximations to $\Pi_k(x)$ and $\pi_k(x)$. Estimates are also known for a related quantity that we have not discussed, the number of $n \leq x$ such that $\omega(n) = k$. See [Ten, chapter II.6].

Exercises

1 Show that the statement $p_n \sim n \log n$ implies that $\pi(x) \sim x/\log x$ as $x \to \infty$.

2 Let $p_1, p_2, \ldots, p_m$ be the primes not greater than $x^{1/2}$. Show that

$$\pi_2(x) = \pi\left(\frac{x}{p_1}\right) + \cdots + \pi\left(\frac{x}{p_m}\right) - \tfrac{1}{2}m(m+1).$$

3 With c_n defined as in the text, show that c is the k-fold convolution of u_P with itself.

4 Let f, g be strictly positive functions, and let $F(x) = \int_2^x f(t)\,dt$ (and $G(x)$ similarly). Assume (or prove) that if $f(x) \sim g(x)$ and $F(x) \to \infty$ as $x \to \infty$, then $F(x) \sim G(x)$ as $x \to \infty$. Now let

$$S_1(x) = \sum_{p \in P[x]} \frac{\log p}{p^{1/2}}, \qquad S_2(x) = \sum_{p \in P[x]} \frac{1}{p^{1/2}}.$$

Starting from either $\theta(x) \sim x$ or $\pi(x) \sim \mathrm{li}(x)$, prove that $S_1(x) \sim 2x^{1/2}$ and

$$S_2(x) \sim \int_2^x \frac{1}{t^{1/2}\log t}\,dt \qquad \text{as } x \to \infty.$$

(*Note that by exercise 6 of section 1.5, this integral* $\sim 2x^{1/2}/\log x$.)

5 Revisit exercise 3 of section 2.6 to show that (with the same notation), for any $\varepsilon > 0$, we have $\omega(n) \leq (1+\varepsilon)g(n)$ for all large enough n.

4

Prime numbers in residue classes: Dirichlet's theorem

The number $30n + r$ is a multiple of 2, 3 or 5 unless r is one of 1, 7, 11, 13, 17, 19, 23, 29. So all prime numbers (apart from 2, 3, 5) are in one of the eight residue classes $\{30n + r : n \in \mathbb{Z}\}$ corresponding to these values of r. Do they prefer some classes to others? Among numbers from 1 to 3600 (comprising 120 blocks of 30 consecutive numbers), the primes are distributed between the classes as follows:

1	7	11	13	17	19	23	29
62	63	62	62	64	60	62	65

(The table in appendix G displays the primes up to 2520, classified in this way.) The distribution appears to be very nearly equal – but is this a hand-picked example?

Of course, there is nothing special about 30: it can be replaced by a general integer k, and the possible classes are then represented by the numbers r such that $(r, k) = 1$. In this chapter, we shall prove a famous theorem stating that, asymptotically, the prime numbers are indeed equally distributed between these classes. The theorem was essentially discovered by P. G. Dirichlet in 1837, no less than 59 years before the proof of the prime number theorem itself. Dirichlet actually proved that the series $\sum(1/p)$, taken over the primes in any particular residue class, diverges (so that there are infinitely many such primes), but when his ideas are combined with those used to prove the prime number theorem, they deliver the result stated.

The method depends on a branch of mathematics that we have not used so far, group theory. For a fixed k, the residue classes of numbers coprime to k form a group. This is how group theory enters into number theory, and it provides exactly the right concept to reflect the symmetry between these classes. We only need quite basic group theory, and only abelian groups. The key concept for Dirichlet's method is that of *characters* of abelian groups, which we describe in section 4.1, assuming no previous knowledge. *Dirichlet characters*, for each particular k, are essentially the characters of the group just mentioned. For us, they form an interesting new class of arithmetic functions. The corresponding Dirichlet series are called L-functions. The desired theorem is then obtained as another case of our fundamental theorem, with the L-functions playing a part analogous to the role of the zeta function in the proof of the ordinary prime number theorem.

Dirichlet's theorem, and the route to it, amount to a highly elegant and worthwhile extension of the proof of the prime number theorem. The reader is about to enter an especially beautiful area of the mathematical landscape!

4.1 Characters of finite abelian groups

Let G be an abelian group. We write the group operation like multiplication, so that the "product" of elements a, b is ab, and the identity of G is denoted by e. Let $\mathbb{T}$ be the group of complex numbers $\{z : |z| = 1\}$ under multiplication. The *characters* of G are the group homomorphisms χ from G to $\mathbb{T}$. The set of all such characters is denoted by $\mathrm{char}(G)$. As a homomorphism, a character χ satisfies:

(a) $\chi(ab) = \chi(a)\chi(b)$ for $a, b \in G$,

(b) $\chi(e) = 1$,

(c) $\chi(a^{-1}) = \chi(a)^{-1} = \overline{\chi(a)}$ for $a \in G$.

Of course, (a) is the definition of a homomorphism, and properties (b) and (c) follow. The *trivial* character χ_0 is defined by $\chi_0(g) = 1$ for all $g \in G$.

Proposition 4.1.1 *The set* $\mathrm{char}(G)$ *forms an abelian group under multiplication:* $\chi_1\chi_2(a)$ *is defined to be* $\chi_1(a)\chi_2(a)$.

Proof Clearly, $\chi_1\chi_2$ maps into $\mathbb{T}$. It is a homomorphism, since

$$(\chi_1\chi_2)(ab) = \chi_1(a)\chi_1(b)\chi_2(a)\chi_2(b) = (\chi_1\chi_2)(a)(\chi_1\chi_2)(b).$$

Multiplication is associative and commutative. The trivial character χ_0 is clearly an identity in $\mathrm{char}(G)$, and every character χ has an inverse, defined by $\chi^{-1}(a) = \overline{\chi(a)}$. □

We assume from now on that G is finite. Note that if G has n elements, then for all $a \in G$ we have $a^n = e$, hence $\chi(a)^n = 1$. In other words, the possible values of χ are restricted to the complex nth roots of unity.

Proposition 4.1.2 *For every character $\chi \neq \chi_0$, we have*

$$\sum_{a \in G} \chi(a) = 0.$$

Proof Denote this sum by S. Choose $b \in G$ with $\chi(b) \neq 1$. Then $\{ab : a \in G\} = G$, so

$$S = \sum_{a \in G} \chi(ab) = \chi(b) \sum_{a \in G} \chi(a) = \chi(b)S.$$

Hence $S = 0$. □

We deduce that characters are orthogonal, in the following sense:

Corollary 4.1.3 *If χ_1, χ_2 are distinct characters on G, then*

$$\sum_{a \in G} \chi_1(a)\overline{\chi_2(a)} = 0.$$

Proof Note that $\chi_1(a)\overline{\chi_2(a)} = (\chi_1\chi_2^{-1})(a)$. □

For *cyclic* groups, it is easy to describe the characters explicitly:

Proposition 4.1.4 *Let G be a cyclic group of order n, generated by the element a. Write $\omega = \exp(2\pi i/n)$. Then $\text{char}(G)$ is also a cyclic group of order n, generated by χ_1, where $\chi_1(a^k) = \omega^k$ for $1 \leq k \leq n$.*

Proof Let χ be a character, and let $\chi(a) = z$. Since $a^n = e$, we must have $z^n = 1$. Further, $\chi(a^k) = z^k$ for each k. The possible choices of z are $\omega, \omega^2, \ldots, \omega^n (= 1)$. These give characters $\chi_1, \chi_2, \ldots, \chi_n$ $(= \chi_0)$, and clearly $\chi_r = \chi_1^r$. □

Proposition 4.1.5 *Let G be a finite abelian group and let H be a subgroup. Then any character χ on H can be extended to give a character χ_1 on G.*

Proof Choose an element a of $G \setminus H$. We will show that χ can be extended to a character defined on the subgroup $\langle H, a\rangle$ generated by H and a. If this subgroup is not the whole of G, we then choose another element b outside $\langle H, a\rangle$ and repeat the process. After finitely many repetitions, we will have extended χ to the whole of G.

Let a^m be the lowest power of a that belongs to H. Then $\chi(a^m)$ is already defined, and clearly we must define $\chi_1(a)$ to be a complex number ω such that $\omega^m = \chi(a^m)$. We choose any such ω. Now elements of $\langle H, a\rangle$ are of the form $g = ha^r$, where $h \in H$ and $r \in \mathbb{Z}$, and we would like to define $\chi_1(g)$

to be $\chi(h)\omega^r$. However, this representation of g is not unique: suppose that $h'a^s$ is another such representation. Then $a^{s-r} = hh'^{-1} \in H$, from which it follows that $s - r = km$ for some k. Hence $h = h'a^{km}$, and since χ is a character on H, we have

$$\chi(h) = \chi(h')\omega^{km} = \chi(h')\omega^{s-r},$$

or $\chi(h)\omega^r = \chi(h')\omega^s$. This shows that it is consistent to define $\chi_1(g)$ in the way stated. It is now easy to verify that χ_1 is a character on $\langle H, a\rangle$. For if $g_1 = h_1 a^r$ and $g_2 = h_2 a^s$ are two elements, then

$$\begin{aligned} \chi_1(g_1 g_2) &= \chi_1(h_1 h_2 a^{r+s}) \\ &= \chi(h_1 h_2)\omega^{r+s} \\ &= \chi(h_1)\omega^r \chi(h_2)\omega^s \\ &= \chi_1(g_1)\chi_1(g_2). \end{aligned}$$

□

Corollary 4.1.6 *Let G be a finite abelian group, and let a,b be distinct members of G. Then there is a character χ on G such that $\chi(a) \neq \chi(b)$.*

Proof Let $c = ab^{-1}$, and let the element c have order m. Let H be the cyclic subgroup generated by c. As in 4.1.4, define a character on H with $\chi(c) = \omega$, where $\omega = e^{2\pi i/m}$. Since $a = bc$, we have $\chi(a) = \omega\chi(b) \neq \chi(b)$. If $H \neq G$, use 4.1.5 to extend χ to a character on G. □

We now describe a very neat duality between G and char(G). Since char(G) is an abelian group, it has its own characters. Furthermore, the elements of G itself provide examples of such characters in a natural way.

Proposition 4.1.7 *Given $a \in G$, a corresponding character $\hat{a}$ is defined on* char(G) *by setting $\hat{a}(\chi) = \chi(a)$. Further, distinct elements a,b of G give rise to distinct characters $\hat{a}, \hat{b}$.*

Proof The function $\hat{a}$ maps into $\mathbb{T}$. Also, it is a homomorphism, because

$$\hat{a}(\chi_1\chi_2) = (\chi_1\chi_2)(a) = \chi_1(a)\chi_2(a) = \hat{a}(\chi_1)\hat{a}(\chi_2).$$

By 4.1.6, if $a \neq b$, then $\hat{a} \neq \hat{b}$. □

This reversal gives the following dual version of 4.1.2 and 4.1.3:

Proposition 4.1.8 *Let G be a finite abelian group. Then:*

(i) *if $a \neq e$, then* $\sum_{\chi\in\text{char}(G)} \chi(a) = 0$;
(ii) *if $a \neq b$, then* $\sum_{\chi\in\text{char}(G)} \chi(a)\overline{\chi(b)} = 0$.

Proof Apply 4.1.2 and 4.1.3 to $\hat{a}$, $\hat{b}$. □

Of course, $\sum_{\chi \in \mathrm{char}(G)} \chi(e)$ is simply the number of characters. Using orthogonality again, we can now show that this number is the same as the order of G.

Proposition 4.1.9 *If G is any finite abelian group, then* $\mathrm{char}(G)$ *has the same number of elements as G.*

Proof Let G consist of n elements $a_1, a_2, \dots a_n$, and let $\mathrm{char}(G)$ have m elements. For each character χ, define a corresponding element of $\mathbb{C}^n$ by

$$A(\chi) = [\chi(a_1), \chi(a_2), \dots, \chi(a_n)].$$

By 4.1.3, these are m mutually orthogonal elements of $\mathbb{C}^n$. But such a set of elements can have at most n members. Hence $m \le n$.

Now let k be the number of characters on $\mathrm{char}(G)$. By what we have just proved, $k \le m$. But 4.1.7 shows that $k \ge n$. Hence in fact $k = m = n$. □

This also shows that $\{\hat{a} : a \in G\}$ is the full set of characters on $\mathrm{char}(G)$.

A standard result of group theory states that every finite abelian group is a direct product of cyclic groups (this can be proved by a more determined application of the step-by-step process in the proof of 4.1.5). Given this result, it follows quite easily that $\mathrm{char}(G)$ is actually isomorphic to G (which of course implies 4.1.9). However, this fact is not important for our purposes. Of course, 4.1.4 proves it for cyclic groups.

We emphasize that these results apply to *abelian* groups. For general groups, the notion "character" is defined in a different way, which simplifies to our definition when the group is abelian. Non-abelian groups can fail to have any non-trivial characters in the sense of our definiton.

Real characters. A real character is a homomorphism to the two-element group $\{1, -1\}$. Clearly, a character χ is real if and only if $\chi^2 = \chi_0$. The following results give some insight into the nature of real characters.

Proposition 4.1.10 *Let G be a finite abelian group. There are only non-trivial real characters on G if G has even order, $2n$. If χ is such a character, and $H = \{a : \chi(a) = 1\}$, then H is a subgroup of order n.*

Conversely, if H is any subgroup of order n, then a real character χ is defined by

$$\chi(a) = \begin{cases} 1 & \text{if } a \in H, \\ -1 & \text{if } a \in G \setminus H. \end{cases}$$

Proof Let χ be a non-trivial real character, and let $H = \{a : \chi(a) = 1\}$ (the kernel of χ). Then H is a subgroup. Let b be an element with $\chi(b) = -1$. If c is another such element, then $c = ab$ for some $a \in G$, and clearly $\chi(a) = 1$, so $a \in H$. Hence the set of elements c with $\chi(c) = -1$ is exactly the coset Hb, and G is the union of H and Hb. So G must have even order, $2n$, and H has order n.

Now let H be any such subgroup, and let χ be defined as stated. Choose an element b of $G \setminus H$. Then $G \setminus H$ is the coset Hb. We have to show that $\chi(a_1a_2) = \chi(a_1)\chi(a_2)$ for all a_1, a_2 in G. This is obvious if $a_1, a_2 \in H$. Suppose that $a_1 \in H$ and $a_2 \in Hb$, say $a_2 = hb$. Then $a_1a_2 = (a_1h)b$, so $\chi(a_1a_2) = -1 = \chi(a_1)\chi(a_2)$. Finally, suppose that a_1 and a_2 are in Hb, say $a_1 = h_1b$ and $a_2 = h_2b$. If a_1a_2 were in Hb (say equal to h_3b), we would have $h_1h_2b^2 = h_3b$, and hence $b = h_1^{-1}h_2^{-1}h_3 \in H$, a contradiction. So $a_1a_2 \in H$, and again the desired equality holds. □

Obviously, if χ is a real character, then $\chi(a^2) = 1$ for all $a \in G$. The set of squares $\{a^2 : a \in G\}$ is a subgroup of G, which we denote by $\text{sq}(G)$. Hence we have:

Proposition 4.1.11 *Let G be an abelian group of order $2n$. If* $\text{sq}(G)$ *has n elements, then the only non-trivial real character on G is as described in 4.1.10, with $H = \text{sq}(G)$.* □

Exactly this situation arises in many of the number-theoretic groups considered in the next section.

Exercises

1. Show that the mapping $a \to \hat{a}$ (defined in 4.1.7) is itself a group homomorphism.

2. Show that $\text{char}(\mathbb{Z})$ is isomorphic to the group $\mathbb{T}$.

3. Let G be a group of even order. By considering pairs χ, χ^{-1}, show that the number of real characters on G (including χ_0) is even, and hence that there is at least one non-trivial real character.

4. Let $G \times H$ denote the direct product of abelian groups G and H. Show that $\text{char}(G \times H)$ is isomorphic to $\text{char}(G)\times$ $\text{char}(H)$.

5. Let D_n be the group $\{1, -1\}^n$, in other words, the set of all n-tuples $\varepsilon = (\varepsilon_1, \varepsilon_2, \ldots, \varepsilon_n)$ with $\varepsilon_j \in \{1, -1\}$ for each j, and multiplication

defined by $(\delta\varepsilon)_j = \delta_j\varepsilon_j$. Show that the characters on D_n are of the form

$$\chi_S(\varepsilon) = \prod_{j \in S} \varepsilon_j$$

for all subsets S of $\{1, 2, \ldots, n\}$.

6 Develop the method of 4.1.5 to give an alternative proof of 4.1.9.

4.2 Dirichlet characters

Fix an integer $k \geq 2$. For every integer r, we denote by $\hat{r}$ (or $r\hat{}$ when r is a longer expression) the set of all integers congruent to r mod k. In other words,

$$\hat{r} = \{nk + r : n \in \mathbb{Z}\} = \{\ldots,\ r - k,\ r,\ r + k,\ r + 2k,\ \ldots\}.$$

The distinct such sets are $\hat{0}, \hat{1}, \ldots, (k-1)\hat{}$. They are called the *residue classes* modulo k, and between them they form a partition of $\mathbb{Z}$. Note that $\hat{r} = \hat{s}$ if and only if $r \equiv s \pmod{k}$ (which in turn is equivalent to $s \in \hat{r}$). The reader may be familiar with the fact (trivial once one has absorbed the basic idea) that the residue classes form a group under "addition mod k": one defines $\hat{r} + \hat{s}$ to be $(r + s)\hat{}$.

Here we shall be concerned with the fact that a suitable subset of the residue classes form a group under *multiplication* mod k. This fact is not quite so trivial: it embodies some results of elementary number theory. Write

$$E_k = \{r : 1 \leq r \leq k - 1 \text{ and } (r, k) = 1\},$$

$$G_k = \{\hat{r} : r \in E_k\}.$$

Although these sets are part of the basic currency of number theory, there is no really established notation for them. However, G_k is called the "reduced set of residue classes" mod k. By definition, the number of members of E_k (and G_k) is Euler's $\phi(k)$. If k is prime, then $E_k = \{1, 2, \ldots, k - 1\}$ and $\phi(k) = k - 1$, while E_{30} has the eight members 1, 7, 11, 13, 17, 19, 23, 29.

Clearly, if r belongs to E_k, then so does $k - r$. By pairing elements, it follows easily that $\phi(k)$ is even for $k \geq 3$, and that $\sum_{r \in E_k} r = \frac{1}{2}k\phi(k)$.

Proposition 4.2.1 *It is consistent to define "multiplication" on G_k by putting $\hat{r}\hat{s} = (rs)\hat{}$, and this operation makes G_k an abelian group.*

Proof Suppose that $\hat{r}_1 = \hat{r}$ and $\hat{s}_1 = \hat{s}$. Then there are integers a, b such that $r_1 = r + ak$ and $s_1 = s + bk$, so that

$$r_1 s_1 = rs + (as + br + ab)k,$$

hence $(r_1 s_1)\hat{} = (rs)\hat{}$. This shows that the definition is consistent. Also, it defines an element of G_k, since if $(r,k) = (s,k) = 1$, then $(rs,k) = 1$.

The operation is commutative and associative, since ordinary multiplication has these properties. The element $\hat{1}$ is clearly an identity for it. Now choose any element r of E_k. Since $(r,k) = 1$, there exist integers s, t such that $rs + kt = 1$. Then $(s,k) = 1$ (since a common divisor of s and k must be a divisor of 1) and $\hat{r}\hat{s} = (rs)\hat{} = \hat{1}$, so $\hat{s}$ is the inverse of $\hat{r}$. □

Equivalently, one can make E_k itself into a group by defining the "product" of r and s to be the element of E_k that is congruent to rs (mod k). Some notation such as $r \times_k s$ is then needed for this product.

A famous application of 4.2.1 (though not important for our purposes) is the *Euler-Fermat theorem.* In any group G of order n, one has $a^n = e$ for each element a. Hence, for all $r \in E_k$, we have $(\hat{r})^{\phi(k)} = \hat{1}$, in other words, $r^{\phi(k)} \equiv 1 \pmod{k}$.

We shall not need any further general facts about the groups G_k, but some particular examples will appear in the pages that follow.

Now let $\hat{\chi}$ be a character on G_k. We define a corresponding function χ on $\mathbb{Z}$ by:

$$\chi(r) = \begin{cases} \hat{\chi}(\hat{r}) & \text{if } (r,k) = 1, \\ 0 & \text{if } (r,k) > 1. \end{cases}$$

The functions χ on $\mathbb{Z}$ obtained in this way are called *Dirichlet characters mod k.* The set of them will be denoted by $\text{char}(k)$. The *principal* character χ_0 is the one corresponding to the trivial character on G_k: note that it is not entirely "trivial", since it takes the value 1 when $(r,k) = 1$ and 0 when $(r,k) > 1$.

Proposition 4.2.2 *If χ is a Dirichlet character mod k, then*

(i) *χ is completely multiplicative;*

(ii) *χ is periodic, with period k;*

(iii) *$\chi(r) = 0$ whenever $(r,k) > 1$.*

Conversely, let χ be a function on $\mathbb{Z}$, with values in $\mathbb{T} \cup \{0\}$. If χ has properties (i), (ii) and (iii), then it is a Dirichlet character mod k.

Proof Suppose that χ is a Dirichlet character. From the way it was defined, it is clear that (ii) and (iii) hold. Choose integers r, s. If $(r,k) > 1$ or $(s,k) > 1$, then $(rs,k) > 1$, and $\chi(rs) = \chi(r)\chi(s) = 0$. If $(r,k) = (s,k) = 1$, then $(rs)\hat{} = \hat{r}\hat{s}$, and hence

$$\chi(rs) = \hat{\chi}(\hat{r}\hat{s}) = \hat{\chi}(\hat{r})\hat{\chi}(\hat{s}) = \chi(r)\chi(s).$$

Now suppose that conditions (i), (ii) and (iii) hold. Then it is consistent to define $\hat{\chi}$ on G_k by putting $\hat{\chi}(\hat{r}) = \chi(r)$, and in the same way we have $\hat{\chi}(\hat{r}\hat{s}) = \hat{\chi}(\hat{r})\hat{\chi}(\hat{s})$. Then $\hat{\chi}$ is a character on G_k, and χ is the corresponding Dirichlet character. □

Example 1 Let $k = 4$. Then $G_4 = \{\hat{1}, \hat{3}\}$, with $(\hat{3})^2 = \hat{1}$. The two characters are given by $\chi_0(3) = 1$ (the principal character) and $\chi_1(3) = -1$. Both characters take the value 0 at every even integer, and clearly $\chi_0(2n+1) = 1$ and $\chi_1(2n+1) = (-1)^n$.

The orthogonality property of group characters translates as follows for Dirichlet characters.

Proposition 4.2.3 *If χ is a non-principal Dirichlet character mod k, then*

$$\sum_{r \in E_k} \chi(r) = \sum_{r=1}^{k-1} \chi(r) = \sum_{r=1}^{k} \chi(r) = 0.$$

Proof Clearly, $\sum_{r \in E_k} \chi(r) = \sum_{\hat{r} \in G_k} \hat{\chi}(\hat{r}) = 0$. The other sums stated are the same, since the terms added are all 0. □

Of course, $\sum_{r=1}^{k-1} \chi_0(r)$ equals $\phi(k)$.

Note that $\chi(k-1) = \chi(-1)$ for any $\chi \in \text{char}(k)$. Since $(-1)^2 = 1$ and $\chi(1) = 1$, the value of $\chi(-1)$ is either 1 or -1. We shall say that χ is *even* if $\chi(-1) = 1$ and *odd* if $\chi(-1) = -1$. (Although this terminology is natural, it does not seem to be very firmly established.) Note also that

$$\chi(k-r) = \chi(-r) = \chi(-1)\chi(r).$$

We shall write $S_\chi(x) = \sum_{1 \le r \le x} \chi(r)$ for $x \ge 1$. For non-principal Dirichlet characters, we also write

$$M(\chi) = \sup_{x \ge 1} |S_\chi(x)|.$$

Proposition 4.2.4 *If χ is a non-principal Dirichlet character mod k, then* $|M(\chi)| \le \frac{1}{2}\phi(k)$.

Proof By 4.2.3, $S_\chi(k) = 0$, and hence $S_\chi(k+n) = S_\chi(n)$. So S_χ is periodic with period k, and it is enough to consider n between 1 and $k-1$. If $n \leq \frac{1}{2}k$, then there are at most $\frac{1}{2}\phi(k)$ members of E_k between 1 and n, so $|S_\chi(n)| \leq \frac{1}{2}\phi(k)$. Similarly, if $\frac{1}{2}k < n < k$, then $|S_\chi(n)| = |S_\chi(k) - S_\chi(n)| \leq \frac{1}{2}\phi(k)$. □

As we shall see, the quantity $M(\chi)$ is important in various estimations, so a more accurate bound for its value is of interest. *Polya's inequality* states that $|M(\chi)| \leq \sqrt{3}k^{1/2}\log k$ (see [Ap] or [HST]).

We now use the second form of orthogonality (4.1.8) to show how Dirichlet characters relate to the summation of an arithmetic function within separate residue classes. For a fixed integer r and any $x \geq 1$, write

$$S(x,r,k) = \{n : 1 \leq n \leq x \text{ and } n \equiv r \pmod{k}\}.$$

If $a(n)$ is an arithmetic function and χ is a Dirichlet character mod k, we also write

$$A(x,r,k) = \sum_{n \in S(x,r,k)} a(n),$$

$$A_\chi(x) = \sum_{n \leq x} \chi(n)a(n).$$

Proposition 4.2.5 *Let $r \in E_k$. Then, with the above notation,*

$$A(x,r,k) = \frac{1}{\phi(k)} \sum_{\chi \in \mathrm{char}(k)} \overline{\chi}(r)A_\chi(x).$$

Proof For fixed $n \leq x$, we have by 4.1.8:

$$\sum_{\chi \in \mathrm{char}(k)} \overline{\chi}(r)\chi(n) = \begin{cases} 0 & \text{if } n \notin S(x,r,k), \\ \phi(k) & \text{if } n \in S(x,r,k). \end{cases}$$

Hence

$$\begin{aligned} \sum_{\chi \in \mathrm{char}(k)} \overline{\chi}(r)A_\chi(x) &= \sum_{\chi \in \mathrm{char}(k)} \overline{\chi}(r) \sum_{n \leq x} \chi(n)a(n) \\ &= \sum_{n \leq x} a(n) \sum_{\chi \in \mathrm{char}(k)} \overline{\chi}(r)\chi(n) \\ &= \phi(k) \sum_{n \in S(x,r,k)} a(n) \\ &= \phi(k)A(x,r,k). \end{aligned}$$

□

The rest of this section is designed to give the reader more feeling for the nature of Dirichlet characters, but it is not strictly necessary for the theorems of the following sections.

If k is prime, then the group G_k is cyclic. The same is true when $k = 4$ or k is of the form p^n or $2p^n$, where $p \geq 3$ is prime (and only in these cases). A generator of the cyclic group is then called a *primitive root* mod k. For proofs, see books on number theory. Of course, when the group is cyclic, the characters are easily described as in 4.1.4.

Example 2 The group G_5 is generated by $\hat{2}$, since 2, 2^2, 2^3, 2^4 are congruent, respectively, to 2, 4, 3, 1. So a Dirichlet character is fully determined by its value at 2. The four characters are obtained by taking $\chi(2)$ to be 1, i, -1, $-i$ in turn.

The next result gives an easy way to build up Dirichlet characters with a composite modulus.

Proposition 4.2.6 *Let* $k = k_1 k_2$. *If* χ_1, χ_2 *are Dirichlet characters mod* k_1, k_2, *respectively, then a Dirichlet character mod* k *is defined by* $\chi(r) = \chi_1(r)\chi_2(r)$.

Proof It is clear that χ is completely multiplicative and has period k. Also, if $(r, k) > 1$, then either $(r, k_1) > 1$ or $(r, k_2) > 1$, so $\chi(r) = 0$. The statement follows, by 4.2.2. □

Conversely, if $(k_1, k_2) = 1$, then all Dirichlet characters mod k are of this form (see exercise 10).

We now consider *real* Dirichlet characters. As we saw in 4.1.11, if exactly half the elements of a group G are squares, then the only non-trivial real character on G is the function taking the value 1 at squares and -1 at non-squares.

Proposition 4.2.7 *Let* k *be an odd prime, say* $k = 2q + 1$. *Then exactly* q *elements of* G_k *are squares. Hence there is exactly one non-principal Dirichlet character mod* k, *defined in the way just described.*

Proof If $(\hat{r})^2 = (\hat{s})^2$, then k divides into $r^2 - s^2 = (r+s)(r-s)$. Since k is prime, it divides either $r + s$ or $r - s$. In other words, s is congruent (mod k) to either r or $-r$. Hence $1^2, 2^2, \ldots, q^2$ are all distinct mod k. The integers $q + 1, \ldots, 2q$ can be written (in the opposite order) as $k - 1, \ldots, k - q$, and their squares coincide (mod k) with those already listed. □

As the reader may know, the number r is said to be a *quadratic residue* mod k if $\hat{r}$ is a square in G_k. The real character just described is known as the *Legendre symbol*, usually denoted by $(r \mid k)$.

Example 3 Let $k = 11$. Then 1^2, 2^2, 3^2, 4^2, 5^2 are congruent (mod 11) to 1, 4, 9, 5, 3, respectively. Hence the unique non-principal real character is given by the following (in which we write + for 1 and − for −1):

1	2	3	4	5	6	7	8	9	10
+	−	+	+	+	−	−	−	+	−

To construct a similar table for larger values of k, note that the values of successive squares (mod k) are easily found by addition. However, given the value of $\chi(-1)$ (see exercise 4), it is an amusing recreation to construct the table, evaluating only a few squares and then using the multiplicative and periodic properties together with $\chi(k-r) = \chi(-1)\chi(r)$. (The reader may care to try this for some cases, e.g. $k = 131, 163, 191$.)

Clearly, the situation of 4.2.7 occurs whenever G_k is cyclic: the squares are then precisely the even powers of a generator. In other cases, the situation can be quite different, as the next example shows.

Example 4 Consider $G_8 = \{\hat{1}, \hat{3}, \hat{5}, \hat{7}\}$. The square of each element is $\hat{1}$. It follows that all the characters are real. Also, if a, b, c are $\hat{3}$, $\hat{5}$, $\hat{7}$ in any order, then $ab = c$. It is easily deduced that the characters are as follows:

	1	3	5	7
χ_0	+	+	+	+
χ_1	+	+	−	−
χ_2	+	−	+	−
χ_3	+	−	−	+

Exercises

1. Write out the multiplication table for the group G_7 and identify an element that generates the group.

2. If p is prime and $n \geq 2$, show that any Dirichlet character mod p is also a Dirichlet character mod p^n.

3. Describe the non-principal real characters mod 5, 7 and 13. Also, write out the four real characters mod 20 obtained by multiplying the real characters mod 4 and 5.

4 Let k be an odd prime. By considering pairs $\hat{r}$, $(\hat{r})^{-1}$ in G_k, show that $(-1)\hat{}$ is a square in G_k if $k \equiv 1 \pmod 4$, and not a square if $k \equiv -1 \pmod 4$.

5 Let χ be the non-principal real character mod 191. By considering the values of χ at 100, 200, 192, 196 and 207, find $\chi(2)$, $\chi(3)$, $\chi(5)$ and $\chi(23)$. Deduce $\chi(67)$ and $\chi(43)$. Given that $\chi(-1) = -1$ (see the previous exercise), deduce the values of $\chi(19)$, $\chi(7)$ and $\chi(41)$. If so inclined, continue to determine the complete list of values of χ.

6 Let χ be a non-principal character mod k. Write $A(\chi) = \sum_{r=1}^{k-1} r\chi(r)$. Use Abel summation to show that $A(\chi) = -\sum_{r=1}^{k-1} S_\chi(r)$. Show also that, if χ is even, then $A(\chi) = 0$. [*Hint:* Substitute $k - r$ for r in the sum.]

7 Let χ_0 be the principal character mod k. Write $\rho_k = \phi(k)/k$ and

$$S_0(x) = \sum_{r \le x} \chi_0(r), \qquad A_0(x) = \sum_{r \le x} \frac{\chi_0(r)}{r}.$$

Show that $S_0(x) = \rho_k x + q_1(x)$, where $|q_1(x)| \le \frac{1}{2}\phi(k)$. Use Abel's summation formula to deduce that, for a certain constant L_k, we have

$$A_0(x) = \rho_k \log x + L_k + q_2(x),$$

where $|q_2(x)| \le \phi(k)/x$. Show also that if k is prime, then $L_k = \gamma\rho_k + \frac{1}{k}\log k$.

8 Show that if χ is a Dirichlet character, then $(\chi * \chi)(n) = \chi(n)\tau(n)$ for all n. Let $T_\chi(x) = \sum_{n \le x} \chi(n)\tau(n)$, and let ρ_k, L_k be as in the previous exercise. Adapt the proof of 2.5.1 to show that $T_\chi(x) = O(x^{1/2})$ for non-principal χ, while

$$T_{\chi_0}(x) = \rho_k^2 x \log x + \rho_k(2L_k - \rho_k)x + O(x^{1/2}).$$

Write $T(x, r, k) = \sum_{n \in S(x,r,k)} \tau(n)$. Deduce an estimate for $T(x, r, k)$ when $r \in E_k$, and by subtraction an estimate for $T(x, 0, k)$ when k is prime (and in particular when $k = 2$).

9 Let $p \ge 5$ be a prime, $p = 2q + 1$. Show that $\sum_{j=1}^{q} j^2$ is a multiple of p. Deduce that if $p \equiv -1 \pmod 4$ and χ is the non-principal real character mod p, then $\sum_{r=1}^{p-1} r\chi(r)$ is an odd multiple of p (hence non-zero).

10 Let $k = k_1 k_2$, where $(k_1, k_2) = 1$. Let χ_1, χ_2 be non-principal characters mod k_1, k_2, respectively. Let $\chi(r) = \chi_1(r)\chi_2(r)$ for all r. Use the Chinese remainder theorem to show that χ is not the principal

character mod k. Assuming that $\phi(k) = \phi(k_1)\phi(k_2)$, deduce that all Dirichlet characters mod k are as described in 4.2.6.

11 Let χ be a character mod k. The *Gauss sum* $G(n, \chi)$ is defined by

$$G(n, \chi) = \sum_{r=1}^{k-1} \chi(r) e\left(\frac{rn}{k}\right),$$

where $e(x) = e^{2\pi i x}$. Prove the following identities:

(i) $G(-n, \chi) = \chi(-1)G(n, \chi)$ for all n,

(ii) $G(n, \overline{\chi}) = \chi(-1)\overline{G(n, \chi)}$ for all n,

(iii) $G(n\chi) = \overline{\chi(n)}G(1, \chi)$ for $n \in E_k$.

Deduce from (i) and (ii) that if χ is real, then $G(n, \chi)$ is real if χ is even, and purely imaginary if χ is odd.

4.3 Dirichlet L-functions

Elementary properties

Let χ be a Dirichlet character mod k. The corresponding *Dirichlet L-function* is the function defined by the associated Dirichlet series:

$$L(\chi, s) = \sum_{n=1}^{\infty} \frac{\chi(n)}{n^s}.$$

The notation $L(s, \chi)$ is often used instead of $L(\chi, s)$. However, we shall think of it purely as a function of s (the author would prefer the notation $L_\chi(s)$, but this is not in common use). These functions play a role in the proof of the Dirichlet prime number theorem analogous to the role of the zeta function in the ordinary prime number theorem.

We restate the basic results on Dirichlet series as applied to L-functions. If $\chi = \chi_0$, the principal character, then the series above converges when $\operatorname{Re} s > 1$. If $\chi \neq \chi_0$, then $\chi(n)$ has bounded partial sums: with our previous notation, $|S_\chi(x)| \leq M(\chi)$ for all x, where $S_\chi(x) = \sum_{n \leq x} \chi(n)$. By 1.7.7, it follows that the series for $L(\chi, s)$ converges whenever $\operatorname{Re} s > 0$. In both cases, 1.7.10 shows that $L(\chi, s)$ is a holomorphic function of s where defined, with

$$L'(\chi, s) = -\sum_{n=1}^{\infty} \frac{\chi(n) \log n}{n^s}.$$

Let χ be a Dirichlet character and χ_0 the corresponding principal character. Some obvious facts are

(a) $L(\overline{\chi},\overline{s}) = \overline{L(\chi,s)}$;

(b) for $\sigma > 1$, we have $|L(\chi,s)| \le L(\chi_0,\sigma) \le \zeta(\sigma)$, also $|L(\chi,s) - 1| \le L(\chi_0,\sigma) - 1$.

Restating the formulae and estimations in 1.7.5 and 1.7.7, we obtain:

Proposition 4.3.1 *Let χ be any Dirichlet character and $X \ge 1$. Within the region of convergence, we have*

$$L(\chi,s) = s\int_1^\infty \frac{S_\chi(x)}{x^{s+1}}\,dx,$$

and further

$$L(\chi,s) = \sum_{n\le X}\frac{\chi(n)}{n^s} + r_X(s),$$

where

$$r_X(s) = -\frac{S_\chi(X)}{X^s} + s\int_X^\infty \frac{S_\chi(x)}{x^{s+1}}\,dx.$$

If χ is a non-principal character, and $s = \sigma + it$ with $\sigma > 0$, then

$$|L(\chi,s)| \le \frac{|s|}{\sigma}M(\chi) \qquad \text{and} \qquad |r_X(s)| \le \frac{M(\chi)}{X^\sigma}\left(\frac{|s|}{\sigma}+1\right). \qquad \square$$

Example Let χ be the character mod 4 given by $\chi(1) = 1$, $\chi(3) = -1$. Then, clearly,

$$L(\chi,s) = 1 - \frac{1}{3^s} + \frac{1}{5^s} - \frac{1}{7^s} + \cdots.$$

Note that $L(\chi,1) = \pi/4$.

Estimates in terms of t

Proposition 4.3.2 *Let χ be a non-principal character, with $M(\chi) = M$. Then, for $\sigma \ge 1$, $t \ge 0$,*

$$|L(\chi,\sigma+it)| \le \log[M(t+2)] + 2.$$

Proof We use the final estimate in 4.3.1. With X to be chosen, we have

$$\left|\sum_{n\le X}\frac{\chi(n)}{n^s}\right| \le \sum_{n\le X}\frac{1}{n} \le \log X + 1.$$

Also, $|s|/\sigma \le (\sigma + t)/\sigma \le t + 1$, hence

$$|r_X(s)| \le \frac{M}{X}(t+2).$$

Take $X = M(t+2)$ to obtain the statement. □

In particular, $|L(\chi, 1)| \le \log M(\chi) + 3$. For refinements and other cases, see the exercises.

Next, we give a corresponding estimate for $|L'(\chi, s)|$. (This is not needed if we follow Newman's method in the ensuing proof of Dirichlet's theorem.)

Proposition 4.3.3 *Let χ be a non-principal character, with $M(\chi) = M$. Then, for $\sigma \ge 1$, $t \ge 0$,*

$$|L'(\chi, \sigma + it)| \le \tfrac{1}{2}\left(\log[M(t+2)] + 2\right)^2.$$

Proof If $\sigma > 2$, then $|L'(\chi, s)| \le |\zeta'(\sigma)| < 2$, so the statement holds. So assume that $1 \le \sigma \le 2$. It will be convenient to work with $|s|$ rather than t. Let $f(x) = \log x / x^s$ and $S(x) = \sum_{n \le x} \chi(n)$. Then $-L'(\chi, s) = \sum_{n=1}^{\infty} \chi(n) f(n)$. By 1.3.8, for any $X \ge 1$, this equals

$$\sum_{n \le X} \chi(n) f(n) - S(X) f(X) - \int_X^\infty S(x) f'(x)\, dx.$$

If $\sigma \ge 1$, then, by 3.1.7,

$$\left| \sum_{n \le X} \chi(n) f(n) \right| \le \sum_{n \le X} \frac{\log n}{n} \le \tfrac{1}{2}(\log X)^2 + 1.$$

Now

$$f'(x) = \frac{1}{x^{s+1}}(1 - s \log x),$$

hence

$$\begin{aligned} \int_X^\infty |f'(x)|\, dx &\le \int_X^\infty \frac{1}{x^2}(1 + |s| \log x)\, dx \\ &= \frac{1}{X} + |s| \frac{1 + \log X}{X}, \end{aligned}$$

and so

$$\begin{aligned} \left| S(X) f(X) + \int_X^\infty S(x) f'(x)\, dx \right| &\le \frac{M \log X}{X} + \frac{M}{X} + \frac{M|s|}{X}(1 + \log X) \\ &= \frac{M}{X}(|s| + 1)(\log X + 1). \end{aligned}$$

Now take $X = M(|s|+1)$. We obtain

$$|L'(\chi, s)| \le \tfrac{1}{2}(\log X)^2 + \log X + 2$$
$$\le \tfrac{1}{2}(\log X + 2)^2.$$

The original statement follows, since $X \le M(t+2)$. □

The Euler product and related series

If χ is any Dirichlet character and Re $s > 1$, then $f(n) = \chi(n)/n^s$ is completely multiplicative and $\sum_{n=1}^{\infty} |f(n)|$ is convergent, so, as a direct case of the generalized Euler product (2.1.1), we have:

Theorem 4.3.4 *Let χ be any Dirichlet character. Then, for* Re $s > 1$,

$$L(\chi, s) = \prod_{p \in P} \left(1 - \frac{\chi(p)}{p^s}\right)^{-1}.$$

In particular, $L(\chi, s) \ne 0$. □

Of course, if χ is a character mod k, then the prime divisors of k can be omitted from the product, since $\chi(p) = 0$ for such p.

Corollary 4.3.5 *If χ is a real character, then $L(\chi, \sigma) > 0$ for $\sigma > 1$.*

Proof For each prime p, we have $p^\sigma > 1$, and hence $1 - \chi(p)/p^\sigma > 0$. □

Theorem 4.3.6 *Let χ be any Dirichlet character. Then, for* Re $s > 1$,

$$\frac{1}{L(\chi, s)} = \sum_{n=1}^{\infty} \frac{\chi(n)\mu(n)}{n^s}.$$

Proof 1 Apply 2.2.1, with $f(n)$ as above. □

Proof 2 Since $\mu * u = e_1$ and χ is completely multiplicative, 1.8.9 gives $(\chi\mu) * \chi = \chi e_1 = e_1$. The statement follows. □

Corollary 4.3.7 *Let χ be a Dirichlet character and χ_0 the corresponding principal character. Then, for $s = \sigma + it$ with $\sigma > 1$,*

$$\frac{1}{|L(\chi, s)|} \le L(\chi_0, \sigma), \qquad \left|\frac{1}{L(\chi, s)} - 1\right| \le L(\chi_0, \sigma) - 1.$$ □

Note Although the defining series for $L(\chi, s)$ converges for Re $s > 0$ (when χ is non-principal), it is important to realise that the above results have only been shown to apply when Re $s > 1$.

Theorem 4.3.8 *Let χ be any Dirichlet character. For* $\mathrm{Re} s > 1$, *a logarithm of $L(\chi, s)$ is given by*

$$\sum_{p \in P} \sum_{m=1}^{\infty} \frac{\chi(p)^m}{m p^{ms}}.$$

Proof Similar to the corresponding proof for $\zeta(s)$ (see 2.3.2), noting that a logarithm of $1 - \chi(p)/p^s$ is

$$-\sum_{m=1}^{\infty} \frac{\chi(p)^m}{m p^{ms}}. \qquad \square$$

Theorem 4.3.9 *Let χ be any Dirichlet character. Then, for* $\mathrm{Re}\, s > 1$,

$$-\frac{L'(\chi, s)}{L(\chi, s)} = \sum_{n=1}^{\infty} \frac{\chi(n)\Lambda(n)}{n^s}.$$

Proof 1 By differentiation of the above series for $\log L(\chi, s)$, as in 2.3.4. $\square$

Proof 2 Since $\Lambda * u = \ell$ and χ is completely multiplicative, 1.8.9 gives $(\chi\Lambda) * \chi = \ell\chi$. So, by 1.8.8, we have for $\mathrm{Re}\, s > 1$

$$L(\chi, s) \sum_{n=1}^{\infty} \frac{\chi(n)\Lambda(n)}{n^s} = \sum_{n=1}^{\infty} \frac{\chi(n) \log n}{n^s} = -L'(\chi, s). \qquad \square$$

L-functions for principal characters

As before, write $PD(k)$ for the set of prime divisors of k.

Proposition 4.3.10 *Let χ_0 be the principal character mod k. Then, for* $\mathrm{Re}\, s > 1$,

$$L(\chi_0, s) = \zeta(s) \prod_{p \in PD(k)} \left(1 - \frac{1}{p^s}\right).$$

Proof For all primes p that do not divide into k, we have $(p, k) = 1$, hence $\chi_0(p) = 1$. So, by the Euler product,

$$L(\chi_0, s) = \prod \left\{ \left(1 - \frac{1}{p^s}\right)^{-1} : p \in P,\ p \text{ not a divisor of } k \right\}$$

$$= \zeta(s) \prod_{p \in PD(k)} \left(1 - \frac{1}{p^s}\right). \qquad \square$$

Note In the case when k is prime, this statement is easily seen without the Euler product. For if $k = p$, then $\chi_0(j) = 1$ except when j is a multiple of p, so that

$$L(\chi, s) = \zeta(s) - \sum_{n=1}^{\infty} \frac{1}{(np)^s} = \left(1 - \frac{1}{p^s}\right)\zeta(s).$$

Now let

$$G(s) = \prod_{p \in PD(k)} \left(1 - \frac{1}{p^s}\right).$$

Clearly, this finite product is well-defined and holomorphic for all s. So we can use our extension of $\zeta(s)$ to $\operatorname{Re} s > 0$ (or, using the second version, to $\operatorname{Re} s > -1$) to extend the definition of $L(\chi_0, s)$ to such s: we simply define $L(\chi_0, s) = \zeta(s)G(s)$.

Proposition 4.3.11 *Let χ_0 be the principal character mod k. Then, for any $\delta > 0$, there are positive constants m_δ and M_δ (depending on k) such that, for all $s \neq 1$ with* $\operatorname{Re} s \geq \delta$,

$$m_\delta |\zeta(s)| \leq |L(\chi_0, s)| \leq M_\delta |\zeta(s)|.$$

Further, $M_1 \leq \log k + 1$ and $m_1 \geq (\log 2)/(\log k + 1)$.

Proof Let the prime divisors of k be $p_1, \ldots, p_r$ (so that, in our usual notation, $r = \omega(k)$). If $\operatorname{Re} s \geq \delta$, then $|p_j^s| \geq p_j^\delta$, so the statement holds with

$$m_\delta = \prod_{j=1}^{r} \left(1 - \frac{1}{p_j^\delta}\right), \qquad M_\delta = \prod_{j=1}^{r} \left(1 + \frac{1}{p_j^\delta}\right).$$

To estimate these quantities when $\delta = 1$, note that $2^r \leq p_1 p_2 \ldots p_r \leq k$, so that $r \leq \log k / \log 2$. Since $p_j \geq j + 1$, we have

$$M_1 = \prod_{j=1}^{r} \left(1 + \frac{1}{p_j}\right) \leq \prod_{j=1}^{r} \left(1 + \frac{1}{j+1}\right) = \prod_{j=1}^{r} \frac{j+2}{j+1} = \tfrac{1}{2}(r+2) < \log k + 1,$$

and similarly

$$m_1 = \prod_{j=1}^{r} \left(1 - \frac{1}{p_j}\right) \geq \frac{1}{r+1} > \frac{\log 2}{\log k + 1}. \qquad \square$$

Hence, in particular, $L(\chi_0, 1 + it) \neq 0$ for all $t \neq 0$, and upper and lower estimates for $|L(\chi_0, 1+it)|$ follow from those for the zeta function. (An upper estimate more like the one in 4.3.2 can be found by a different method: see exercise 6).

An exact expression for m_1 can be given in terms of Euler's ϕ-function: indeed, we have by 2.2.15:

$$m_1 = G(1) = \prod_{j=1}^{r} \left(1 - \frac{1}{p_j}\right) = \frac{\phi(k)}{k}.$$

Proposition 4.3.12 *Let χ_0 be the principal character mod k. Then*

$$\lim_{s \to 1} (s-1) L(\chi_0, s) = \frac{\phi(k)}{k}.$$

Hence $L(\chi_0, s)$ has a simple pole at 1, with residue $\phi(k)/k$.

Proof With $G(s)$ as above, we have

$$\lim_{s \to 1} (s-1) G(s) \zeta(s) = G(1) \lim_{s \to 1} (s-1) \zeta(s) = G(1) = \frac{\phi(k)}{k}. \quad \square$$

Non-vanishing of L-functions on Re s = 1

We will establish this property by a suitable modification of the method used for the zeta function.

Proposition 4.3.13 *Let χ be any Dirichlet character and χ_0 the corresponding principal character. Then, for any $\sigma > 1$ and any t (including 0), we have*

$$L(\chi_0, \sigma)^3 |L(\chi, \sigma + it)|^4 |L(\chi^2, \sigma + 2it)| \geq 1.$$

Proof Let

$$F(\sigma + it) = 3 \log L(\chi_0, \sigma) + 4 \log L(\chi, \sigma + it) + \log L(\chi^2, \sigma + 2it),$$

in which the logarithms are those given by the series in 4.3.8. In this series, we can take the summation to be over the set P' of all primes that are not divisors of k. So we have

$$F(\sigma + it) = \sum_{p \in P'} \sum_{m=1}^{\infty} \frac{1}{m p^{m\sigma}} \left(3 + 4\chi(p)^m p^{-imt} + \chi(p)^{2m} p^{-2imt}\right).$$

Write $\chi(p) = e^{i\alpha}$. Then the expression in the bracket is $3 + 4e^{im\theta} + e^{2im\theta}$, where $\theta = \alpha - t \log p$. By the inequality $3 + 4\cos\theta + \cos 2\theta \geq 0$ (see 3.1.9), the real part of this is non-negative. Hence Re $F(\sigma + it) \geq 0$: this is equivalent to the statement. $\square$

Theorem 4.3.14 *If χ is a non-real character, then $L(\chi, 1+it) \neq 0$ for all t. If χ is a real character, then $L(\chi, 1+it) \neq 0$ for all $t \neq 0$.*

Proof By 4.3.11, the statement is true for principal characters, so we assume that χ is not principal. Suppose that $L(\chi, 1+it) = 0$ for some t. Then

$$\frac{L(\chi, \sigma+it)}{\sigma-1} \to L'(\chi, 1+it) \quad \text{as } \sigma \to 1^+.$$

Let

$$\begin{aligned} G(\sigma) &= L(\chi_0, \sigma)^3 L(\chi, \sigma+it)^4 L(\chi^2, \sigma+2it) \\ &= (\sigma-1)\,[(\sigma-1)L(\chi_0,\sigma)]^3 \left(\frac{L(\chi,\sigma+it)}{\sigma-1}\right)^4 L(\chi^2, \sigma+2it). \end{aligned}$$

By 4.3.13, $|G(\sigma)| \geq 1$ for all $\sigma > 1$. By 4.3.12, $(\sigma-1)L(\chi_0, \sigma) \to \phi(k)/k$ as $\sigma \to 1$. Also, under either hypothesis, $L(\chi^2, \sigma+2it)$ tends to the finite limit $L(\chi^2, 1+2it)$ as $\sigma \to 1$. Indeed, if χ is not real, then $\chi^2 \neq \chi_0$, so χ^2 is continuous at all points on the line Re $s = 1$, including 1 itself. If χ is real, then $\chi^2 = \chi_0$, and $L(\chi_0, s)$ (with its extended definition) is continuous on Re $s = 1$ except at 1. Together, these statements imply that $G(\sigma) \to 0$ as $\sigma \to 1^+$, contradicting the inequality $|G(\sigma)| \geq 1$. □

Still following the method used for the zeta function, we can give bounds in this result.

Theorem 4.3.15 *If χ is any non-real character mod k, then, for $\sigma \geq 1$ and $t \geq 0$,*

$$\frac{1}{|L(\chi, \sigma+it)|} \leq 4\{\log[k(t+1)] + 2\}^7.$$

If χ is a non-principal real character, then, for $\sigma \geq 1$, $t \geq 1$,

$$\frac{1}{|L(\chi, \sigma+it)|} \leq 4\{\log[k(t+1)] + 2\}^6 (\log k + 1)(\log t + 5).$$

Proof Given estimates of the form

$$|L(\chi^2, \sigma+2it)| \leq M_1(t),$$

$$|L'(\chi, \sigma+it)| \leq M_2(t),$$

the proof of 3.1.14 applies without change to show that

$$\frac{1}{|L(\chi, \sigma+it)|} \leq 2^5 M_1(t) M_2(t)^3.$$

(Note that $L(\chi,\sigma) \le \zeta(\sigma)$.) Since $M(\chi) \le \frac{1}{2}k$ for non-principal characters mod k, 4.3.3 gives

$$M_2(t) = \tfrac{1}{2}\{\log[k(t+1)] + 2\}^2$$

for any $t \ge 0$. When χ is not real (so that $\chi^2 \ne \chi_0$), 4.3.2 gives

$$M_1(t) = \log\left((2t+2)\frac{k}{2}\right) + 2 = \log[k(t+1)] + 2,$$

while if χ is real (so that $\chi^2 = \chi_0$), 4.3.11 and 3.1.6 give

$$M_1(t) = (\log k + 1)(\log 2t + 4) < (\log k + 1)(\log t + 5)$$

for $t \ge 1$. The stated estimates follow. □

$L(\chi, 1) \ne 0$ *for real characters*

We now address the missing case in 4.3.14: is $L(\chi, 1)$ non-zero for real characters χ? The answer is yes, but a completely different method is needed: the way in which the proof of 4.3.14 fails is more than just a minor difficulty. Several different proofs are known, all non-trivial. Here we give a proof that provides a non-zero lower estimate without any extra effort.

Lemma 4.3.16 *Let χ be a real character. Then, for all n,*

$$(\chi * u)(n) \ge 0 \qquad \text{and} \qquad (\chi * u)(n^2) \ge 1.$$

Proof Since χ and u are multiplicative, so is $\chi * u$, by 1.8.8. Hence it is enough to show that if p is prime, then $(\chi * u)(p^n) \ge 0$ for all n, and $(\chi * u)(p^n) \ge 1$ if n is even. Now

$$(\chi * u)(p^n) = 1 + \sum_{r=1}^{n} \chi(p)^r.$$

Hence $(\chi * u)(p^n)$ equals $n+1$ if $\chi(p) = 1$ and 1 if $\chi(p) = 0$, while if $\chi(p) = -1$, then

$$(\chi * u)(p^n) = \begin{cases} 0 & \text{if } n \text{ is odd} \\ 1 & \text{if } n \text{ is even.} \end{cases}$$

So the required statement holds in each case. □

Lemma 4.3.17 *Let*

$$g(x) = \frac{1}{x} - \frac{1}{e^x - 1}.$$

Then $g(x)$ is decreasing and $0 < g(x) < \frac{1}{2}$ for $x > 0$.

Proof We have $g(x) = u(x)/v(x)$, where

$$u(x) = e^x - 1 - x = \sum_{n=2}^{\infty} \frac{x^n}{n!}, \qquad v(x) = x(e^x - 1) = \sum_{n=2}^{\infty} \frac{x^n}{(n-1)!}.$$

Since $n! \geq 2(n-1)!$ for all $n \geq 2$, it follows that $u(x) \leq \frac{1}{2}v(x)$, hence $g(x) \leq \frac{1}{2}$, for all $x > 0$. To show that $g(x)$ is decreasing, one can consider the derivative, but the following, which applies to a general ratio of two series, is more revealing. Suppose that $u(x) = \sum_{n=0}^{\infty} a_n x^n$ and $v(x) = \sum_{n=0}^{\infty} b_n x^n$, in which $a_n, b_n > 0$ and $a_n = c_n b_n$ with c_n decreasing (in our case, we have $c_n = 1/(n+2)$ after cancelling x^2). Let $v_n(x) = \sum_{r=0}^{n} b_r x^r$. By Abel summation,

$$u(x) = \sum_{n=0}^{\infty} c_n b_n x^n = \sum_{n=0}^{\infty} (c_n - c_{n+1}) v_n(x).$$

The statement will follow if we can show that $v_n(x)/v(x)$ is decreasing or, equivalently, that $v(x)/v_n(x)$ is increasing, for each n. Now

$$\frac{v(x)}{v_n(x)} = 1 + \sum_{s=n+1}^{\infty} b_s \frac{x^s}{v_n(x)}$$

and $v_n(x)/x^s = \sum_{r=0}^{n} b_r x^{r-s}$, which is clearly decreasing for $s \geq n+1$. So $x^s/v_n(x)$, and hence also $v(x)/v_n(x)$, is increasing. □

Theorem 4.3.18 *If χ is a non-principal real character, then $L(\chi, 1) > 0$. In fact,*

$$L(\chi, 1) \geq \frac{\pi}{8M(\chi) + 16}.$$

Proof For $\alpha > 0$, let

$$F(\alpha) = \sum_{n=1}^{\infty} (\chi * u)(n) e^{-\alpha n}.$$

By 4.3.16 and integral comparison for series,

$$\begin{aligned} F(\alpha) \geq \sum_{n=1}^{\infty} e^{-\alpha n^2} &= \sum_{n=0}^{\infty} e^{-\alpha n^2} - 1 \\ &\geq \int_0^{\infty} e^{-\alpha x^2}\, dx - 1 \\ &= \frac{1}{2}\left(\frac{\pi}{\alpha}\right)^{1/2} - 1. \end{aligned} \tag{4.1}$$

On the other hand, we may reverse summation and write $n = mj$ to obtain

$$F(\alpha) = \sum_{n=1}^{\infty} e^{-\alpha n} \sum_{j|n} \chi(j) = \sum_{j=1}^{\infty} \chi(j) \sum_{m=1}^{\infty} e^{-\alpha mj}.$$

By the geometric series, we now have

$$\begin{aligned} F(\alpha) &= \sum_{j=1}^{\infty} \chi(j) \frac{1}{e^{\alpha j} - 1} \\ &= \sum_{j=1}^{\infty} \frac{\chi(j)}{\alpha j} - \sum_{j=1}^{\infty} \chi(j) h(j) \\ &= \frac{1}{\alpha} L(\chi, 1) - \sum_{j=1}^{\infty} \chi(j) h(j), \end{aligned} \tag{4.2}$$

where $h(x) = g(\alpha x)$, with $g(x)$ defined as in 4.3.17. Write $M(\chi) = M$. Since $h(x)$ is decreasing, discrete Abel summation (in the form 1.3.4) gives

$$\left| \sum_{j=1}^{\infty} \chi(j) h(j) \right| \leq M h(1) = M g(\alpha) \leq \tfrac{1}{2} M.$$

So, by (4.1) and (4.2),

$$\frac{1}{\alpha} L(\chi, 1) \geq \tfrac{1}{2} \left(\frac{\pi}{\alpha} \right)^{1/2} - 1 - \tfrac{1}{2} M,$$

or

$$L(\chi, 1) \geq a \alpha^{1/2} - b \alpha,$$

where $a = \frac{1}{2} \pi^{1/2}$, $b = 1 + \frac{1}{2} M$. The right-hand side is maximised by taking $\alpha^{1/2} = a/2b$, which gives

$$L(\chi, 1) \geq \frac{a^2}{4b} = \frac{\pi}{8M + 16}. \qquad \square$$

Note Recall that Liouville's function λ is completely multiplicative and has $\lambda(p) = -1$ for every prime p. By 2.2.13, $\lambda * u = u_S$, where S is the set of squares. So λ has the property in 4.3.16, in the most minimal form. However, by 3.4.6, $\sum_{n=1}^{\infty} \lambda(n)/n = 0$, in contrast to 4.3.18. Given this fact, the conclusion from the proof of 4.3.18 is that the summation function of $\lambda(n)$ is not bounded. (Actually, the method shows much more than this: see exercise 7).

The *generalized Riemann hypothesis* is the conjecture that all Dirichlet L-functions are non-zero for Re $s > \frac{1}{2}$. Like the ordinary Riemann hypothesis, this problem has resisted all attempts at solution for more than a century. A variant is the conjecture that all L-functions derived from non-principal real characters are strictly positive for all $\sigma > 0$. This can be shown fairly easily for a large number of particular characters (see exercise 8), but again a general proof is elusive.

Exercises

1 Let χ be a non-principal character, with $M(\chi) = M$. Use the integral expression for $L(\chi, \sigma)$ and the fact that $|S_\chi(x)|$ is bounded by x (as well as by M) to show that

$$|L(\chi,\sigma)| \leq \begin{cases} \log M + 1 & \text{for } \sigma = 1, \\ (M^{1-\sigma} - \sigma)/(1-\sigma) & \text{for } \sigma \neq 1. \end{cases}$$

By writing $1/x^\sigma \leq M^{1-\sigma}/x$ on $[1, M]$, obtain the alternative bound $M^{1-\sigma}(\sigma \log M + 1)$ for $0 < \sigma \leq 1$.

2 Let $M(\chi) = M$. If $M \geq 3$, show by the method of exercise 1 that

$$|L'(\chi, 1)| \leq \tfrac{1}{2}(\log M)^2 + 1.$$

3 Let χ be a non-principal character, with $M(\chi) = M$. Modify the method of 4.3.2 to prove that, for $0 < \sigma < 1$ and $t \geq 0$,

$$|L(\chi, \sigma + it)| \leq \frac{1}{\sigma(1-\sigma)}[M(t+2)]^{1-\sigma}.$$

4 Let χ be a non-principal character. Show that $\chi * \chi = \tau\chi$. Apply an exercise from section 1.8 to show that the series $\sum_{n=1}^{\infty} \tau(n)\chi(n)/n^s$ converges whenever Re $s > \frac{1}{2}$. (By the uniqueness principle for analytic functions, it follows that the sum is $L(\chi, s)^2$.)

5 Let χ be a real character. Deduce from 4.3.16, or from the Euler product, that $L(\chi, \sigma) \geq \zeta(2\sigma)/\zeta(\sigma)$ for $\sigma > 1$.

6 Recall that $\zeta(s, a) = \sum_{n=0}^{\infty} 1/(n+a)^s$ for Re $s > 1$ and $a > 0$ (section 3.1, exercise 10). By collecting together the terms $n = qk + r$ for each r, show that

$$L(\chi, s) = \frac{1}{k^s} \sum_{r=1}^{k-1} \chi(r)\zeta\left(s, \frac{r}{k}\right)$$

for Re $s > 1$. Now use the extension of $\zeta(s, a)$ by Euler's summation formula to extend $L(\chi, s)$ to Re $s > -1$ (you may assume that this coincides with previous definitions where they apply). Deduce that, for non-principal χ,

$$L(\chi,0) = -\frac{1}{k}\sum_{r=1}^{k-1} r\chi(r)$$

and that, for any χ, when $\sigma \geq 1$ and $t \geq 2$,

$$|L(\chi,\sigma+it)| \leq \log k + \log t + 5$$

7 Let λ be Liouville's function, and let $S_\lambda(x) = \sum_{n\leq x}\lambda(n)$. With $g(x)$ and $h(x)$ defined as in the proof of 4.3.18, show that

$$\sum_{n=1}^{\infty} \lambda(n)h(n) = -\sum_{n=1}^{\infty} e^{-\alpha n^2}.$$

Use the fact that $g(x)$ is bounded above by both $\frac{1}{2}$ and $1/x$ to show that $\int_0^\infty x^{-1/2}g(x)\,dx \leq 2\sqrt{2}$. Now choose $X > 0$ and suppose that $-S_\lambda(x) \leq Mx^{1/2}$ for all $x \geq X$. Use Abel's summaton formula to show that

$$-\sum_{n=1}^{\infty} \lambda(n)h(n) \leq X + M(2/\alpha)^{1/2}.$$

Deduce that $M \geq \pi^{1/2}/2\sqrt{2}$, and hence that, for any $c < \pi^{1/2}/2\sqrt{2}$, there exists arbitrarily large x with $S_\lambda(x) < -cx^{1/2}$.

8 Let χ be a non-principal real character mod k. Let $S_1(n) = \sum_{j=1}^{n}\chi(j)$ and $S_2(n) = \sum_{j=1}^{n} S_1(j)$. Suppose that $S_2(n) \geq 0$ for $1 \leq n \leq k$. Suppose also that the function f is decreasing and *strictly convex*, so that $f(\frac{1}{2}x+\frac{1}{2}y) < \frac{1}{2}f(x)+\frac{1}{2}f(y)$ when $x < y$. By performing discrete Abel summation twice, show that $\sum_{r=1}^{k}\chi(r)f(r) > 0$. Deduce that $L(\chi,\sigma) > 0$ for all $\sigma > 0$. (*Note*: This condition is satisfied by many real characters, including those mod p for all primes $p \leq 41$. The reader may care to verify this for some cases.)

9 Write $\rho_k = \phi(k)/k$. Show that, for $|s-1| < 1$,

$$L(\chi_0, s) = \frac{\rho_k}{s-1} + \alpha_0 + \alpha_1(s-1) + \cdots,$$

where

$$\alpha_0 = \gamma\rho_k + \rho_k \sum_{p\in PD(k)} \frac{\log p}{p-1}.$$

(Note that, by 3.4.4, this is the value of the limit L_k in exercise 7 of section 4.2.)

4.4 Prime numbers in residue classes

We are now ready to prove the theorem stated in the introduction to this chapter, that the prime numbers, asymptotically, are distributed equally between residue classes. This is another application of our fundamental theorems, with the Dirichlet L-functions playing the part of the zeta function.

Throughout this section, $k \geq 3$ is a fixed integer. Write

$$\psi_\chi(x) = \sum_{n \leq x} \chi(n)\Lambda(n),$$

where Λ is von Mangoldt's function. We now describe the asymptotic behaviour of $\psi_\chi(x)$, separately for principal and non-principal characters.

Proposition 4.4.1 *Let χ_0 be the principal character mod k. Then*

$$\psi_{\chi_0}(x) \sim x \quad \text{as } x \to \infty.$$

Proof We deduce this from the corresponding statement for $\psi(x)$. Now

$$\psi(x) - \psi_{\chi_0}(x) = \sum_{n \leq x} [1 - \chi_0(n)]\Lambda(n).$$

The only non-zero terms in this sum are those given by $n = p^m$, where p is a prime divisor of k. Let these divisors be $p_1, \ldots, p_s$, and for each j, let m_j be the largest m such that $p_j^m \leq x$. The contribution to the above sum of the powers of p_j is $m_j \log p_j$, which is not greater than $\log x$. Hence

$$\psi(x) - \psi_{\chi_0}(x) \leq s \log x,$$

and the statement follows. □

Our results on L-functions have cleared the way for the following theorem, which is the heart of the matter.

Theorem 4.4.2 *If χ is any non-principal character mod k, then*

$$\frac{\psi_\chi(x)}{x} \to 0 \quad \text{as } x \to \infty.$$

Proof We apply the fundamental theorem 3.4.2. Let

$$f(s) = -\frac{L'(\chi, s)}{L(\chi, s)}.$$

The series

$$\sum_{n=1}^{\infty} \frac{\chi(n)\Lambda(n)}{n^s}$$

is absolutely convergent to $f(s)$ when Re $s > 1$. By 4.3.14 and 4.3.18, $L(\chi, s) \neq 0$, hence $f(s)$ is holomorphic, on a region including Re $s \geq 1$. There is no pole at 1, so (with previous notation) $\alpha = 0$ and $\alpha_0 = f(1)$. Condition (FT3) is satisfied, since, by 4.3.3 and 4.3.15, there are constants C, K (depending on k) such that

$$|f(\sigma \pm it)| \leq C[\log(t+2) + K]^9$$

when $\sigma \geq 1$ and $t \geq 1$. Alternatively, condition (FT3$'$) is ensured by Chebyshev's estimate $\psi(x) \leq 2x$. Finally, $|\chi(n)\Lambda(n)| \leq \Lambda(n)$, which (with $g(s) = -\zeta'(s)/\zeta(s)$) satisfies conditions (FT1), (FT2) and (FT3). The statement follows. □

Note At the same time, from the integral and series versions of the fundamental theorems, we have:

$$\int_1^\infty \frac{\psi_\chi(x)}{x^2}\,dx = \sum_{n=1}^\infty \frac{\chi(n)\Lambda(n)}{n} = -\frac{L'(\chi,1)}{L(\chi,1)}.$$

Following the pattern of previous notation, we write

$$S(x,r,k) = \{n : 1 \leq n \leq x \text{ and } n \equiv r \pmod{k}\},$$

$$P(x,r,k) = \{p : p \text{ prime and } p \in S(x,r,k)\},$$

$$\pi(x,r,k) = \text{the number of members of } P(x,r,k),$$

$$\psi(x,r,k) = \sum_{n \in S(x,r,k)} \Lambda(n),$$

$$\theta(x,r,k) = \sum_{p \in P(x,r,k)} \log p.$$

Combining the above results, we now have easily:

Theorem 4.4.3 *Let* $(r,k) = 1$. *Then*

$$\psi(x,r,k) \sim \frac{x}{\phi(k)} \quad \textit{as } x \to \infty,$$

and similarly for $\theta(x,r,k)$.

Proof By 4.4.1 and 4.4.2,

$$\sum_{\chi \in \text{char}(k)} \overline{\chi}(r)\psi_\chi(x) \sim x \quad \text{as } x \to \infty.$$

The statement for $\psi(x,r,k)$ follows, by 4.2.5. Clearly,

$$0 \le \psi(x,r,k) - \theta(x,r,k) \le \psi(x) - \theta(x),$$

which, by 2.4.4, is not greater than $6x^{1/2}$. The statement for $\theta(x,r,k)$ follows. □

The expected statement about $\pi(x,r,k)$ now follows in the same way as the corresponding step (1.6.2) linking $\pi(x)$ and $\theta(x)$.

Theorem 4.4.4 *Let* $(r,k)=1$. *Then*

$$\pi(x,r,k) \sim \frac{1}{\phi(k)} \operatorname{li}(x) \quad \textit{as } x \to \infty.$$

So, asymptotically, the prime numbers are distributed equally between the reduced residue classes mod k.

Proof Clearly, $\theta(x,r,k) = \sum_{n \le x} b_r(n)$, where

$$b_r(n) = \begin{cases} \log n & \text{if } n \text{ is prime and } n \equiv r \pmod{k}, \\ 0 & \text{otherwise.} \end{cases}$$

By Abel's summation formula,

$$\begin{aligned} \pi(x,r,k) &= \sum_{2 \le n \le x} \frac{b_r(n)}{\log n} \\ &= \frac{\theta(x,r,k)}{\log x} + \int_2^x \frac{\theta(t,r,k)}{t(\log t)^2}\, dt. \end{aligned}$$

Meanwhile, we recall that

$$\operatorname{li}(x) = \frac{x}{\log x} - \alpha + \int_2^\infty \frac{t}{(\log t)^2}\, dt,$$

where $\alpha = 2/\log 2$. Let $\varepsilon > 0$ be given, and let x_0 be such that, for $x \ge x_0$,

$$\theta(x,r,k) \le (1+\varepsilon)\frac{x}{\phi(k)}.$$

Then, for $x \ge x_0$,

$$\begin{aligned} \pi(x,r,k) &\le \frac{1+\varepsilon}{\phi(k)} \frac{x}{\log x} + \frac{1+\varepsilon}{\phi(k)} \int_2^x \frac{t}{(\log t)^2}\, dt + C \\ &= \frac{1+\varepsilon}{\phi(k)} (\operatorname{li}(x) + \alpha) + C, \end{aligned}$$

where

$$C = \int_2^{x_0} \left(\frac{\theta(t,r,k)}{t(\log t)^2} - \frac{1+\varepsilon}{(\log t)^2} \right) dt.$$

Hence, for all large enough x, we have

$$\pi(x,r,k) \le (1+2\varepsilon)\frac{\mathrm{li}(x)}{\phi(k)}.$$

Similarly, for large enough x,

$$\pi(x,r,k) \ge (1-2\varepsilon)\frac{\mathrm{li}(x)}{\phi(k)}. \qquad \square$$

Note 1 The only primes not included in one of the sets $S(x,r,k)$ (with $(r,k)=1$) are the prime divisors of k.

Note 2 In particular, there are infinitely many primes in each residue class $\hat{r}$: this is Dirichlet's original theorem.

As another application of the fundamental theorems, we derive the series result analogous to $\sum_{n=1}^{\infty} \mu(n)/n = 0$.

Proposition 4.4.5 *Let χ be a Dirichlet character. Then*

$$\sum_{n=1}^{\infty} \frac{\chi(n)\mu(n)}{n} = \begin{cases} 1/L(\chi,1) & \text{if } \chi \ne \chi_0, \\ 0 & \text{if } \chi = \chi_0. \end{cases}$$

Proof Let $f(s) = 1/L(\chi,s)$. Then $f(s)$ is holomorphic on a region including $\mathrm{Re}\, s \ge 1$, with no pole at 1 (and taking the value 0 there if $\chi = \chi_0$). When $\mathrm{Re}\, s > 1$, it is given by the absolutely convergent Dirichlet series

$$\sum_{n=1}^{\infty} \frac{\chi(n)\mu(n)}{n^s}.$$

Clearly, $|\chi(n)\mu(n)| \le 1$ for all n. For $\sigma \ge 1$, $|f(\sigma \pm it)|$ satisfies the appropriate inequality from 4.3.15. The statement follows, by 3.4.2. $\square$

Of course, this amounts to saying that the series for $1/L(\chi,s)$ is valid at $s=1$.

Example Let χ be the character mod 4 given by $\chi(1)=1$, $\chi(3)=-1$. Then $\chi(2n+1)=(-1)^n$ and $L(\chi,1)=\pi/4$. Hence

$$\sum_{n=0}^{\infty} (-1)^n \frac{\mu(2n+1)}{2n+1} = \frac{4}{\pi}.$$

Exercises

1 Let χ be a non-principal real character. Let

$$\psi^+(x,\chi) = \sum\{\Lambda(n) : n \leq x,\ \chi(n) = 1\},$$

and let $\pi^+(x,\chi)$ be the number of primes $p \leq x$ with $\chi(p) = 1$. Show that

$$\psi^+(x,\chi) \sim \tfrac{1}{2}x \qquad \text{and} \qquad \pi^+(x,\chi) \sim \tfrac{1}{2}\,\mathrm{li}(x) \quad \text{as } x \to \infty.$$

2 Show that for any non-principal character χ,

$$\sum_{p \in P[x]} \chi(p) = o[\pi(x)] \quad \text{as } x \to \infty.$$

3 Let

$$M(x,r,k) = \sum_{n \in S(x,r,k)} \mu(n).$$

Show that if $(r,k) = 1$, then $M(x,r,k) = o(x)$ as $x \to \infty$.

4 Let χ be a non-principal character. From the series version of the main theorem, show that the series

$$\sum_{n \geq 2} \frac{\chi(n)\Lambda(n)}{n \log n} \qquad \text{and} \qquad \sum_{p \in P} \frac{\chi(p)}{p}$$

are convergent. Deduce that for $r \in E_k$,

$$\sum_{p \in P(x,r,k)} \frac{1}{p} = \frac{1}{\phi(k)} \log\log x + O(1).$$

5 Let χ be a non-principal real character, and let $A(x) = \sum_{n \leq x} (\chi * u)(n)$. Prove that

$$\frac{A(x)}{x} \to L(\chi, 1) \quad \text{as } x \to \infty.$$

5

Error estimates and the Riemann hypothesis

No statement about approximation or limits is really complete without an estimate of the difference between the quantities concerned. For example, the statement that $(1 + x/n)^n$ tends to e^x as $n \to \infty$ is much less informative than the statement that it lies between $(1 - x^2/n)e^x$ and e^x. Our other number-theoretic estimations in sections 2.5 and 2.6 all came complete with an error estimate, but so far we have no such estimate for the prime number theorem itself.

In this chapter, we will establish that, for certain constants K, c,

$$|\pi(x) - \mathrm{li}(x)| \leq Kx \exp[-c(\log x)^{1/2}].$$

The ungainly expression on the right is $o[x/(\log x)^k]$ for every $k > 0$, but not $O(x^\alpha)$ for any $\alpha < 1$. This estimate was already obtained by de la Vallée Poussin in his original proof of the prime number theorem and, despite enormous interest in the problem, it has not been seriously improved in the intervening century.

As with our original fundamental theorems, we will actually prove this result in the context of general Dirichlet series, and it applies equally to our second main example, $M(x)$. The method is a fairly straightforward adaptation of our original proof by Mellin inversion (one can also enter it from Newman's method, at the cost of an extra step). The previous (FT3) has to be replaced by a stronger hypothesis. For the main particular cases, $\zeta'(s)/\zeta(s)$ and $1/\zeta(s)$, this means that we will need to know that $\zeta(s) \neq 0$ on a region of the form $\sigma \geq 1 - c/\log t$. The proof of this fact depends on a rather different circle of ideas, and we defer it to section 5.3, inviting the reader to take it on trust and read about the error estimate first.

A huge question mark hangs over this work. Massive numerical evidence suggests that both $\pi(x) - \mathrm{li}(x)$ and $M(x)$ are really $O(x^\alpha)$ for all $\alpha > \frac{1}{2}$, which would mean a radical improvement on de la Vallée Poussin's estimate.

It turns out that both of these statements are actually equivalent to the Riemann hypothesis. In other words, the Riemann hypothesis is equivalent to the error term in the prime number theorem being what computation suggests, rather than what we have been able to prove. The importance attached to the hypothesis is a measure of the respect accorded to the prime number theorem in the world of mathematics! We go part way to explaining the equivalence in section 5.2.

5.1 Error estimates

General theorems

We shall prove more precise versions of the fundamental theorems, incorporating error estimates. Recall that the context is a sequence $a(n)$ and a function $f(s)$ such that:

(FT1) *the series* $\sum_{n=1}^{\infty} \frac{a(n)}{n^s}$ *converges absolutely to* $f(s)$ *when* $\operatorname{Re} s > 1$;

(FT2) $f(s) = \frac{\alpha}{s-1} + \alpha_0 + (s-1)h(s)$, *where* h *is differentiable at 1.*

We replace (FT3) by the following pair of conditions, which are tailor-made for the most important particular cases:

(FT3a) *there exist* c_0, M_0 *such that* f *is differentiable on the set* $|t| \geq 2$, $\sigma \geq 1 - c_0/\log|t|$ *and satisfies* $|f(\sigma + it)| \leq M_0(\log|t|)^3$ *there;*

(FT3b) f *is differentiable at points* $\sigma + it$ *(except 1) with* $|t| \leq 2$ *and* $\sigma \geq 1 - c_0/\log 2$.

We shall use the following elementary facts:

$$(\log t)^n \leq \left(\frac{n}{e}\right)^n t \quad \text{for } t \geq 1, \tag{5.1}$$

$$\int_1^{\infty} \frac{(\log t)^n}{t^2}\, dt = n! \tag{5.2}$$

The new version of the integral fundamental theorem, 3.2.5, is as follows.

Theorem 5.1.1 *Suppose that* $a(n)$ *and* $f(s)$ *satisfy conditions (FT1), (FT2), (FT3a) and (FT3b). Let* $A(x) = \sum_{n \leq x} a(n)$. *Then there exist con-*

stants K, c such that

$$\left| \int_x^\infty \frac{A(y) - \alpha y}{y^2} \, dy \right| \le K \exp[-c(\log x)^{1/2}]$$

for all sufficiently large x.

Proof As before, let $\phi(s) = h(s)/s$. In the proof of 3.2.5 (alternatively, in exercise 2 of section 3.3), we proved that

$$\int_x^\infty \frac{A(y) - \alpha y}{y^2} \, dy = \frac{\alpha_0 - \alpha}{x} - \frac{1}{2\pi i} \int_{L_1} x^{s-1} \phi(s) \, ds.$$

With a small adjustment to M_0, we may assume that (FT3a) holds with $f(s)$ replaced by $(s-1)h(s)$. Since $|s(s-1)| \ge t^2$, this implies that for s as described in (FT3a),

$$|\phi(s)| \le M_0 \frac{(\log |t|)^3}{t^2}. \tag{5.3}$$

Also, $|\phi(\sigma + it)|$ is bounded, say by B, on the rectangle $1 - c_0/\log 2 \le \sigma \le 1$, $|t| \le 2$.

For a certain $T > 2$, to be specified later, denote by I_1, I_2, I_3 the integrals of $x^{s-1}\phi(s)$ on $[1-iT, 1+iT]$, $[1+iT, 1+i\infty)$ and $(1-i\infty, 1-iT]$, respectively. By (5.3),

$$|I_r| \le M_0 \int_T^\infty \frac{(\log t)^3}{t^2} \, dt \qquad (r = 2, 3).$$

By (5.1), $(\log t)^3 \le \rho t^{1/2}$, where $\rho = (6/e)^3$, so that

$$|I_r| \le M_0 \rho \int_T^\infty \frac{1}{t^{3/2}} \, dt = \frac{2M_0\rho}{T^{1/2}} \qquad (r = 2, 3).$$

Now consider I_1. Let $a = 1 - c_0/\log T$. By Cauchy's theorem, we can express I_1 as $J_1 + J_2 + J_3$, where J_1, J_2, J_3 are the integrals on $[a - iT, a + iT]$, $[a + iT, 1 + iT]$, $[1 - iT, a - iT]$, respectively. Note that these lines are within the region where conditions (FT3a) and (FT3b) are satisfied. The point of this is to move further left, introducing (in J_1, the main part) the factor $|x^{s-1}| = x^{a-1}$.

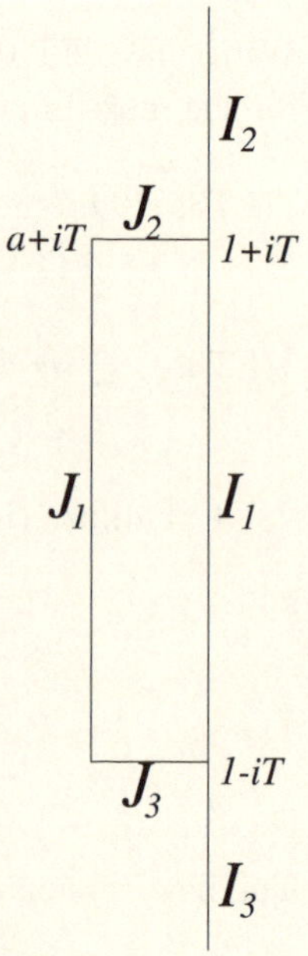

I_2 J_2 $a+iT$ $1+iT$ J_1 I_1 J_3 $1-iT$ I_3

Divide J_1 further, as follows: let J_1' be the integral on $[a - 2i, a + 2i]$ and

J_1'' the integral on $[a+2i, a+iT]$. Then $|J_1'| \le 4x^{a-1}B$, while, by (5.2) and (5.3),

$$|J_1''| \le M_0 x^{a-1} \int_2^\infty \frac{(\log t)^3}{t^2}\, dt \le 6M_0 x^{a-1}$$

(and similarly for $[a-it, a-2i]$). Hence

$$|J_1| \le (12M_0 + 4B)x^{a-1}.$$

We now consider J_2 and J_3. On the intervals in question, $|x^{s-1}| \le 1$, so, for $r = 2, 3$,

$$|J_r| \le M_0(1-a)\frac{(\log T)^3}{T^2} = M_0 c_0 \frac{(\log T)^2}{T^2} \le M_0 c_0 \frac{4}{e^2}\frac{1}{T}.$$

For a given x, we now choose T by

$$\log T = 2c(\log x)^{1/2},$$

where $2c^2 = c_0$. (To ensure that $T > 2$, we must take x greater than a certain value.) Then

$$(a-1)\log x = -c_0\frac{\log x}{\log T} = -c(\log x)^{1/2} = -\tfrac{1}{2}\log T,$$

so $x^{a-1} = T^{-\frac{1}{2}}$, and each I_r satisfies an estimate of the form

$$|I_r| \le K_r \exp[-c(\log x)^{1/2}].$$

Of course, the same is true for the term $(\alpha_0 - \alpha)/x$. □

Note 1 The relation between c and c_0 is $c = (c_0/2)^{1/2}$. In the specific cases considered below, the value of c_0 will be quite small.

Note 2 For this proof, there was no need for the original move of the line of integration from L_c to L_1 in 3.2.5. With small modifications to the above proof, one can take the right-hand line to be L_b, where $b = 1 + 1/\log x$.

The expression $\exp[-c(\log x)^{1/2}]$ will appear in all our error estimates. The next result will help to give a feeling for what it means.

Proposition 5.1.2 *Let* $r(x) = \exp[-c(\log x)^{1/2}]$, *where* $c > 0$. *Then:*

(i) *for any* $\varepsilon > 0$, *we have* $x^\varepsilon r(x) \to \infty$ *as* $x \to \infty$ *(so* $r(x)$ *is ultimately large compared with* $x^{-\varepsilon}$*);*

(ii) *for every* $k \ge 1$, *we have* $r(x) = o[(\log x)^{-k}]$ *as* $x \to \infty$.

Proof (i) This follows from

$$\log[x^{\varepsilon} r(x)] = \varepsilon \log x - c(\log x)^{1/2} \to \infty \quad \text{as } x \to \infty.$$

(ii) Let $r_1(y) = \exp(-cy^{1/2})$. Then

$$\log[y^k r_1(y)] = k \log y - cy^{1/2} \to -\infty \quad \text{as } y \to \infty,$$

so that $y^k r_1(y) \to 0$. The substitution $y = \log x$ gives the statement. □

To translate error estimates for integrals into ones for limits and series, we need a more precise form of the Tauberian lemma 3.4.1. Fortunately, this requires nothing more than a closer look at the original proof. It will be enough to deal with the case when $A(x)$ is increasing and $\alpha = 1$. We require the following lemma.

Lemma 5.1.3 *If* $0 < \delta \leq \frac{1}{2}$, *then:*

$$\delta - \log(1+\delta) > \tfrac{1}{3}\delta^2, \qquad \delta + \log(1-\delta) < -\tfrac{1}{2}\delta^2.$$

Proof From the series for $\log(1+\delta)$, we have

$$\delta - \log(1+\delta) = \left(\frac{\delta^2}{2} - \frac{\delta^3}{3}\right) + \left(\frac{\delta^4}{4} - \frac{\delta^5}{5}\right) + \cdots.$$

Each bracket is positive, so

$$\delta - \log(1+\delta) > \tfrac{1}{2}\delta^2 - \tfrac{1}{3}\delta^3 = \delta^2\left(\tfrac{1}{2} - \tfrac{1}{3}\delta\right) \geq \delta^2\left(\tfrac{1}{2} - \tfrac{1}{6}\right) = \tfrac{1}{3}\delta^2.$$

The second statement follows at once from the series for $\log(1-\delta)$. □

Proposition 5.1.4 *Suppose that* $A(x)$ *is a positive, increasing real function on* $[1,\infty)$ *and that*

$$\left|\int_x^{\infty} \frac{A(y) - y}{y^2}\, dy\right| \leq r(x)$$

for all $x \geq 1$, *where* $r(x)$ *is decreasing. Suppose that* $r(x) < \frac{1}{24}$ *for all* $x \geq a$. *Then*

$$\left|\frac{A(x)}{x} - 1\right| \leq \sqrt{6}\, r(x/2)^{1/2}$$

for all $x \geq 2a$.

Proof Choose $x_0 > 2a$ and suppose that $A(x_0) = (1+\delta_0)x_0$, where $\delta_0 > 0$, so that $\delta_0 = A(x_0)/x_0 - 1$. Let $\delta = \min(\delta_0, \frac{1}{2})$. Let $x_1 = (1+\delta)x_0$, and consider

$$I_1 = \int_{x_0}^{x_1} \frac{A(y) - y}{y^2}\, dy.$$

By our hypothesis, $I_1 \leq 2r(x_0)$. However, as shown in the proof of 3.4.1,

$$I_1 \geq \delta - \log(1+\delta) \geq \tfrac{1}{3}\delta^2.$$

Therefore $\delta^2 \leq 6r(x_0)$. Since $r(x_0) < \frac{1}{24}$, we must have $\delta < \frac{1}{2}$; hence in fact $\delta_0 = \delta$ and the estimation applies to δ_0.

Now suppose that $A(x_0) = (1-\delta_0)x_0$. Define δ as before. Let $x_2 = (1-\delta)x_0$, and consider

$$I_2 = \int_{x_2}^{x_0} \frac{A(y) - y}{y^2}\,dy.$$

Then $|I_2| \leq 2r(x_2) \leq 2r(x_0/2)$ (since $x_2 \geq \frac{1}{2}x_0 \geq a$), and we find

$$I_2 \leq \delta + \log(1-\delta) < -\tfrac{1}{2}\delta^2,$$

so that $\delta^2 \leq 4r(x_0/2)$; hence again $\delta < \frac{1}{2}$ and $\delta_0 = \delta$. □

We can now present our more precise version of 3.4.2, the limit and series forms of the fundamental theorem.

Theorem 5.1.5 *Suppose that $a(n) \geq 0$ for all n, and that $a(n)$ and $f(s)$ satisfy conditions (FT1), (FT2), (FT3a) and (FT3b), with $\alpha > 0$. Let $A(x) = \sum_{n \leq x} a(n)$. Then there exist constants K, c such that, for all sufficiently large x, each of the following quantities is bounded (in modulus) by $K \exp[-c(\log x)^{1/2}]$:*

$$\frac{A(x)}{x} - \alpha, \qquad \sum_{n \leq x} \frac{a(n)}{n} - \alpha \log x - \alpha_0, \qquad \sum_{n > x} \frac{a(n) - \alpha}{n}.$$

Proof We may assume, by taking a scalar multiple, that $\alpha = 1$. The statement for $A(x)/x$ follows at once from 5.1.1 and 5.1.4, with a new K and c; for example, since $\log(x/2) \geq \frac{1}{2}\log x$ for $x \geq 4$, we replace c by $c/(2\sqrt{2})$.

By Abel's summation formula, as in 3.4.2,

$$\sum_{n \leq x} \frac{a(n)}{n} - \alpha \log x - \alpha_0 = F(x) + G(x),$$

where

$$F(x) = \frac{A(x)}{x} - \alpha,$$

$$G(x) = \int_1^x \frac{A(y) - \alpha y}{y^2}\,dy - (\alpha_0 - \alpha) = \int_x^\infty \frac{A(y) - \alpha y}{y^2}\,dy.$$

The second statement follows. Again as in 3.4.2, we have

$$-\sum_{n>x} \frac{a(n)-\alpha}{n} = \sum_{n\le x} \frac{a(n)-\alpha}{n} - (\alpha_0 - \gamma\alpha)$$

$$= \sum_{n\le x} \frac{a(n)}{n} - \alpha \log x - \alpha_0 \left(\sum_{n\le x} \frac{1}{n} - \log x - \gamma \right).$$

The third statement follows, since the last term is $O(1/x)$. □

Note 1 The statements can be taken as applying for all $x \ge 1$, at the cost of possibly increasing K, since there are certainly constants such that the statements hold for $1 \le x \le x_0$ (on such a bounded interval, the left-hand side is bounded above and the right-hand side is bounded below).

Note 2 We have avoided complication by refraining from stating the version of 5.1.4 and 5.1.5 for the case when $a(n)$ is not positive, but is dominated by another function satisfying the conditions. However, the idea is obvious. For the specific case considered below, it will be quite simple to supply the required step "one-off".

Note 3 The shift from $r(x)$ to $r(x)^{1/2}$ in 5.1.4 is not avoidable. For us, this just had the effect of having to choose a new c. However, it would be a serious loss if we were trying to establish an error estimate of the form $Kx^{-\alpha}$, as implied by the Riemann hypothesis; a different method is needed for such cases.

The prime number theorem and the Möbius function

To apply these theorems to

$$-\frac{\zeta'(s)}{\zeta(s)} = \sum_{n=1}^{\infty} \frac{\Lambda(n)}{n^s} \qquad \text{and} \qquad \frac{1}{\zeta(s)} = \sum_{n=1}^{\infty} \frac{\mu(n)}{n^s},$$

we need to know that these functions satisfy (FT3a) and (FT3b). The proof of (FT3a) requires quite a bit of further work, of a rather different nature, so we defer it to section 5.3 and suggest that the reader assumes it for the moment. With this assumption, we can state without further proof:

Theorem 5.1.6 *Let ψ be Chebyshev's function. There are constants K, c such that, for all large enough x (or with a different K, for all $x \ge 2$), we have*

$$|\psi(x) - x| \le Kx \exp[-c(\log x)^{1/2}].$$ □

We now deduce de la Vallée Poussin's error estimate in the prime number theorem itself. Note first that since (by 2.4.4) $\psi(x) - \theta(x) \leq 6x^{1/2}$, the statement in 5.1.6 holds for $\theta(x)$ as well as $\psi(x)$. We show next, a little more generally, how an upper estimate for $\pi(x) - \text{li}(x)$ follows from one for $\theta(x) - x$. The proof is almost a repeat of the original proof (1.6.2) that $\theta(x) \sim x$ implies $\pi(x) \sim \text{li}(x)$.

Lemma 5.1.7 *Suppose that, for all $x \geq x_1$, we have $|\theta(x) - x| \leq KF(x)$, where $F(x)/(\log x)^2$ is increasing. Then there exists K' such that, for all $x \geq x_1$,*

$$|\pi(x) - \text{li}(x)| \leq K' \frac{F(x)}{\log x}.$$

Proof By the expressions for $\pi(x)$ and $\text{li}(x)$ used in 1.6.2,

$$\begin{aligned}\pi(x) - \text{li}(x) &= \alpha + \frac{\theta(x) - x}{\log x} + \int_2^x \frac{\theta(t) - t}{t(\log t)^2}\, dt \\ &= \alpha + J_1 + J_2 \qquad \text{(say)},\end{aligned}$$

where $\alpha = 2/\log 2$. Take $x > x_1$. Clearly, $|J_1| \leq KF(x)/\log x$. Also,

$$J_2 = A + \int_{x_1}^x \frac{\theta(t) - t}{t(\log t)^2}\, dt,$$

where A is the integral on $[x, x_1]$. But for $x_1 \leq t \leq x$, we have

$$\frac{|\theta(t) - t|}{(\log t)^2} \leq K \frac{F(t)}{(\log t)^2} \leq K \frac{F(x)}{(\log x)^2},$$

so

$$|J_2| \leq A + K \frac{F(x)}{(\log x)^2} \int_{x_1}^x \frac{1}{t}\, dt \leq A + K \frac{F(x)}{\log x}.$$

The statement follows. □

Theorem 5.1.8 *There are constants K, c such that, for all sufficiently large x (or, with a different K, for all $x \geq 2$),*

$$|\pi(x) - \text{li}(x)| \leq Kx \exp[-c(\log x)^{1/2}].$$

Proof With c as before, let $F(x) = x \exp[-c(\log x)^{1/2}]$, and let $G(x) = F(x)/(\log x)^2$. By 5.1.7, we only need to show that $G(x)$ is increasing for all large enough x (the conclusion entitles us to put $F(x)/\log x$ on the right-hand side, but this is still only an estimate of the same form). Now

$$\begin{aligned}\log G(x) &= \log x - c(\log x)^{1/2} - 2\log\log x \\ &= y - cy^{1/2} - 2\log y,\end{aligned}$$

where $y = \log x$. The derivative of this function of y is $1 - \frac{1}{2}cy^{-1/2} - 2y^{-1}$, which is clearly positive when y is large enough. □

Hence certainly $\pi(x) - \mathrm{li}(x) = o[x/(\log x)^k]$ for all $k \geq 1$. By 1.5.3 and 1.5.4, it is now finally clear that in the prime number theorem, $\mathrm{li}(x)$ is a better approximation than $x/(\log x - 1)$, which in turn is better than $x/\log x$.

We now deal with the second main case, $1/\zeta(s)$. This time we include the series statement as well as the integral and limit versions. Since $\mu(n)$ is not always positive, we must use the dominating function $\zeta(s)$, as in 3.4.5.

Theorem 5.1.9 *Let $M(x) = \sum_{n \leq x} \mu(n)$. There are constants K, c such that, for large enough x, each of*

$$\int_x^\infty \frac{M(y)}{y^2}\,dy, \qquad \frac{M(x)}{x}, \qquad \sum_{n>x} \frac{\mu(n)}{n}$$

is bounded by $K\exp[-c(\log x)^{1/2}]$. Similar statements apply with $\mu(n)$ replaced by $\mu(n)n^{-it}$ (with the constants K,c depending on t).

Proof First consider $\mu(n)$. The integral is a case of 5.1.1. To deduce the estimates for the limit and the series, we apply 5.1.1 and 5.1.5 to the function

$$\zeta(s) - \frac{1}{\zeta(s)} = \sum_{n=1}^{\infty} \frac{1 - \mu(n)}{n^s}.$$

Since $1 - \mu(n) \geq 0$, this tells us that

$$\frac{[x] - M(x)}{x} - 1$$

has an exponential bound of the type stated. Since $|[x]/x - 1| \leq 1/x$, the statement about $M(x)/x$ follows. By Abel's summation formula,

$$\sum_{n>x} \frac{\mu(n)}{n} = -\frac{M(x)}{x} + \int_x^\infty \frac{M(y)}{y^2}\,dy,$$

so the third estimate follows from the other two.

Now fix $t_0 > 0$ and let $a(n) = \mu(n)n^{-it_0}$, so that $1/\zeta(s)$ is replaced by $1/\zeta(s + it_0)$. This function satisfies condition (FT3a) with $\log t$ replaced by $\log(t + t_0)$, hence the condition as stated with new constants c_1, M_1. □

Note that since $\sum_{n=1}^{\infty} \mu(n)/n = 0$, we have

$$\sum_{n \leq x} \frac{\mu(n)}{n} = -\sum_{n>x} \frac{\mu(n)}{n}.$$

Dirichlet L-functions also satisfy (FT3a) and (FT3b), so a similar error estimate applies in the theorems of section 4.4 (but with the constants depending on the modulus k).

We now describe two applications of these theorems, in the sense of results that are not themselves statements about error estimates. First, we have always suspected that there are fewer primes in $(x, 2x]$ than in $[1, x]$. The next result makes this precise.

Proposition 5.1.10 *We have*

$$2\pi(x) - \pi(2x) \sim \frac{cx}{(\log x)^2} \quad \textit{as } x \to \infty,$$

where $c = 2\log 2$. *In particular,* $2\pi(x) - \pi(2x) \to \infty$ *as* $x \to \infty$.

Proof We prove a similar statement for $\mathrm{li}(x)$. The result then follows because of the fact that $\pi(x) - \mathrm{li}(x) = o[x/(\log x)^2]$. Recall from 1.5.3 that

$$\mathrm{li}(x) = \frac{x}{\log x} + r(x),$$

where $r(x) \sim [x/(\log x)^2]$ as $x \to \infty$. Hence

$$2\,\mathrm{li}(x) - \mathrm{li}(2x) = \frac{2x}{\log x} - \frac{2x}{\log(2x)} + 2r(x) - r(2x).$$

Now

$$\frac{2x}{\log x} - \frac{2x}{\log x + \log 2} = \frac{2x\log 2}{\log x(\log x + \log 2)} \sim \frac{cx}{(\log x)^2} \quad \text{as } x \to \infty,$$

and $r(2x) \sim [2x/(\log 2x)^2] \sim [2x/(\log x)^2]$, so that $2r(x) - r(2x) = o[x/(\log x)^2]$ as $x \to \infty$. So we have

$$2\,\mathrm{li}(x) - \mathrm{li}(2x) \sim \frac{cx}{(\log x)^2} \quad \text{as } x \to \infty,$$

as required. □

Proposition 5.1.11 *We have*

$$\sum_{n=1}^{\infty} \frac{\mu(n)\log n}{n} = -1.$$

Proof We apply the integral version of the fundamental theorem to the function

$$f(s) = -\frac{d}{ds}\frac{1}{\zeta(s)} = \frac{\zeta'(s)}{\zeta(s)^2}.$$

Then

$$f(s) = \sum_{n=1}^{\infty} \frac{a(n)}{n^s} \quad \text{for } \operatorname{Re} s > 1$$

where $a(n) = \mu(n) \log n$. There is no pole at 1: the expansion is

$$f(s) = -1 + 2\gamma(s-1) + \cdots$$

(see 3.1.5). Our condition (FT3) is satisfied, with $P(t) = C(\log t + 5)^{16}$. (Alternatively, Newman's condition (FT3′) is equivalent to $M(x) = O(x/\log x)$, which we now know from 5.1.9.) So, with $A(x)$ defined as usual, theorem 3.2.5 gives

$$\int_1^{\infty} \frac{A(x)}{x^2}\, dx = -1.$$

We cannot deduce limit and series statements by the method of section 3.4, since the dominating function would be $\zeta'(s)$, which has a double pole at 1. Instead, we use 5.1.9 in the form $M(x)\log x/x \to 0$ as $x \to \infty$. By Abel's summation formula,

$$A(x) = M(x)\log x - \int_1^x \frac{M(y)}{y}\, dy.$$

The fact that $\lim_{x\to\infty} M(x)/x = 0$ implies easily that

$$\frac{1}{x}\int_1^x \frac{M(y)}{y}\, dy \to 0 \quad \text{as } x \to \infty.$$

Since $M(x)\log x/x \to 0$, it follows that $A(x)/x \to 0$. By Abel summation again,

$$\sum_{n \le x} \frac{a(n)}{n} = \frac{A(x)}{x} + \int_1^x \frac{A(t)}{t^2}\, dt,$$

and the statement now follows. □

An alternative method

We now outline a second method for 5.1.1, based on a closer analysis of the rate of convergence in the Riemann-Lebesgue lemma, the final step in 3.2.5. Denote by $L_1(\mathbb{R})$ the set of functions ϕ for which

$$\|\phi\|_1 = \int_{-\infty}^{\infty} |\phi(t)|\, dt$$

is convergent, and denote the Fourier transform by $\hat{\phi}$:

$$\hat{\phi}(\lambda) = \int_{-\infty}^{\infty} e^{-i\lambda t}\phi(t)\, dt.$$

As mentioned in the note after 3.2.4, if ϕ' is in $L_1(\mathbb{R})$ and $\phi(t) \to 0$ as $t \to \pm\infty$, then on integrating by parts we obtain

$$\hat{\phi}(\lambda) = \frac{1}{i\lambda}\int_{-\infty}^{\infty} e^{-i\lambda t}\phi'(t)\, dt = \frac{1}{i\lambda}\hat{\phi'}(\lambda).$$

Hence if for all $n \geq 1$, the nth derivative $\phi^{(n)}$ is in $L_1(\mathbb{R})$ and $\phi^{(n)}(t) \to 0$ as $t \to \pm\infty$, then, for $\lambda \neq 0$, we have

$$\hat{\phi}(\lambda) = \frac{1}{(i\lambda)^n}\widehat{\phi^{(n)}}(\lambda), \qquad |\hat{\phi}(\lambda)| \leq \frac{1}{|\lambda|^n}\|\phi^{(n)}\|_1.$$

Hence, for the function $\phi(s)$ used in 3.2.5, we need an estimate of $|\phi^{(n)}(s)|$. Such an estimate can be found quite easily (without worrying about the complications of this function) using the following well-known result from complex analysis.

Let f be differentiable on a region including $\{s : |s - s_0| \leq r\}$, with power series $f(s) = \sum_{n=1}^{\infty} a_n(s - s_0)^n$ on this disc. Suppose that $|f(s)| \leq M$ when $|s - s_0| = r$. Then $|a_n| \leq M/r^n$, so $|f^{(n)}(s_0)| \leq n!M/r^n$ for all n.

Lemma 5.1.12 *Suppose that f satisfies conditions (FT1), (FT2), (FT3a) (now assumed to hold for $|t| \geq \frac{7}{4}$) and (FT3b). Let $\phi(s) = h(s)/s$. Then there exist constants c, M, B such that, for all $t \geq 2$,*

$$|\phi^{(n)}(1 \pm it)| \leq M\frac{n!}{c^n}\frac{(\log t)^{n+3}}{t^2},$$

while for $|t| \leq 2$,

$$|\phi^{(n)}(1 + it)| \leq \frac{Bn!}{c^n}.$$

Hence if $\phi_1(t) = \phi(1 + it)$, then there exists K such that, for all n,

$$\|\phi_1^{(n)}\|_1 \leq K\frac{n!(n+3)!}{c^n}.$$

Proof With a small adjustment to M_0, we may assume that (FT3a) holds with $f(s)$ replaced by $(s - 1)h(s)$. We assume that $c_0 \leq \frac{1}{4}\log 2$. Fix $t_0 \geq 2$. We will use the result just quoted to estimate $|\phi^{(n)}(1 + it_0)|$. One verifies easily that $\log(t + \frac{1}{4})/\log t$ decreases with t, so if we put $\delta = (\log \frac{9}{4})/(\log 2)$, then $\log(t_0 + \frac{1}{4}) \leq \delta \log t_0$. Take $r = c_0/\delta \log t_0$; note that $r \leq \frac{1}{4}$. Let

$s = \sigma + it$ be such that $|s - (1 + it_0)| = r$. Then $t \leq t_0 + \frac{1}{4}$; hence $\log t \leq \delta \log t_0$, and $\sigma \geq 1 - r \geq 1 - c_0/\log t$. So, by (FT3a),

$$|(s-1)h(s)| \leq M_0(\log t)^3 \leq M_0\delta^3(\log t_0)^3.$$

Also, $|s(s-1)| \geq t^2 \geq (\frac{7}{8})^2 t_0^2$, so for a suitable M we have $|\phi(s)| \leq M(\log t_0)^3/t_0^2$. By the result quoted,

$$|\phi^{(n)}(1 + it_0)| \leq M \frac{n!}{r^n} \frac{(\log t_0)^3}{t_0^2} = M \frac{n!}{c^n} \frac{(\log t_0)^3}{t_0^2},$$

where $c = c_0/\delta$.

The function ϕ is continuous, hence bounded (say by B), on the set defined by $1 - c \leq \sigma \leq 1 + c$, $|t| \leq \frac{9}{4}$ (note that $c < c_0$). Now let $0 \leq t_0 \leq 2$. Then $|\phi(s)| \leq B$ for s such that $|s - (1 + it_0)| = c$. The second statement follows.

By (5.2), we now have

$$\int_0^\infty |\phi^{(n)}(1 + it)| \, dt \leq M \frac{n!(n+3)!}{c^n} + 2B \frac{n!}{c^n},$$

and similarly on $(-\infty, 0]$. □

Theorem 5.1.1 now follows by applying the next lemma with $\lambda = \log x$.

Lemma 5.1.13 *Suppose that ϕ_1 satisfies the conclusion of 5.1.12. Then, for all sufficiently large λ,*

$$|\hat{\phi}_1(\lambda)| \leq K e^7 \exp[-(c\lambda)^{1/2}].$$

Proof Let n be the integer such that $(n+3)^2 \leq c\lambda < (n+4)^2$. By comparison with the integral of the concave function log, one has

$$\sum_{r=1}^{n-1} \log r + \tfrac{1}{2} \log n \leq \int_1^n \log x \, dx = n \log n - n + 1,$$

hence $n! \leq e n^{n+\frac{1}{2}} e^{-n}$ (this is a weak form of Stirling's formula). Therefore

$$\begin{aligned} n!(n+3)! &\leq e^2 n^{n+\frac{1}{2}} (n+3)^{n+3+\frac{1}{2}} e^{-2n-3} \\ &\leq e^7 (n+3)^{2n+4} e^{-2n-8} \\ &\leq e^7 (c\lambda)^{n+2} \exp[-2(c\lambda)^{1/2}]. \end{aligned}$$

Hence

$$|\hat{\phi}_1(\lambda)| \leq K \frac{n!(n+3)!}{(c\lambda)^n} \leq K e^7 (c\lambda)^2 \exp[-2(c\lambda)^{1/2}].$$

The statement follows, since $x^4 e^{-2x} \leq e^{-x}$ for large enough x. □

Note 1 Without 5.1.13, this method leads directly to error estimates of the form $C_n/(\log x)^n$ for each n.

Note 2 A more simple-minded variant of 5.1.13 is: *If* $\|\phi_1^{(n)}\|_1 \leq n!/c^n$ *for all* n, *then*

$$|\hat{\phi}_1(\lambda)| \leq e^2(c\lambda)^{1/2}e^{-c\lambda} \quad \textit{for } \lambda > 1/c.$$

(Proof: Take n such that $n \leq c\lambda < n+1$.) If this were to hold for $f(s) = 1/\zeta(s)$, with $c = \frac{1}{2}$, then we would have

$$\left|\int_x^\infty \frac{M(y)}{y^2}\,dy\right| \leq K\frac{\log x}{x^{1/2}}.$$

As we shall see in the next section, this would imply the celebrated Riemann hypothesis. All this would work if we could remove either the $n!$ or the $(n+3)!$ in the last formula in 5.1.12.

Exercises

1 Prove that the integrals

$$\int_1^\infty \frac{|M(x)|}{x^2}\,dx \quad \text{and} \quad \int_1^\infty \frac{|\psi(x)-x|}{x^2}\,dx$$

are convergent.

2 Suppose that $|\theta(x)-x| \leq F(x)$ for all $x \geq 2$, where $F(x)$ is increasing. Show that, for certain constants a and b,

$$|\pi(x) - \mathrm{li}(x)| \leq 3\frac{F(x)}{\log x} + aF(x^{1/2}) + b.$$

3 Integrate $\int_2^x \mathrm{li}(t)/t\,dt$ by parts. Use the result to show that if $|\pi(x) - \mathrm{li}(x)| \leq G(x)$ for all $x \geq 2$, where $G(x)$ is increasing, then

$$|\theta(x) - x| \leq 2G(x)\log x + 2.$$

4 (*Mertens revisited*) Prove that for a certain constant C_1,

$$\sum_{p \in P[x]} \frac{1}{p} = \log\log x + C_1 + r(x),$$

where $r(x) = O[(\log x)^{-k}]$ for each $k \geq 1$. Show also that if $\pi(x) - \mathrm{li}(x) = O(x^\alpha)$ for some $\alpha < 1$, then $r(x) = O(x^{\alpha-1})$.

5.2 Connections with the Riemann hypothesis

We introduced the Riemann hypothesis in section 3.1: it is the conjecture, supported by extended computation, that $\zeta(s)$ has no zeros with Re $s > \frac{1}{2}$. In this section, we explain the close connection between this hypothesis and error estimates. We only include some of the easier proofs. All the proofs omitted here can be found, for example, in [Ing].

If the Riemann hypothesis were true, then (for the two main cases) the contour of integration in 5.1.1 could have been moved almost to the line Re $s = \frac{1}{2}$. A modified version of the argument then leads to the following conclusions:

$$\psi(x) - x = O(x^\alpha), \qquad M(x) = O(x^\alpha)$$

for every $\alpha > \frac{1}{2}$ (and hence the same for $\pi(x) - \mathrm{li}(x)$).

Furthermore, these implications are reversible. We give the proof in this direction, since it is quite an easy consequence of the principle of uniqueness for holomorphic functions.

Proposition 5.2.1 *Let* $0 \le \alpha < 1$. *If either* $\psi(x) - x$ *or* $M(x)$ *is* $O(x^\alpha)$ *for* $x > 1$, *then* $\zeta(s) \neq 0$ *whenever* $\alpha <$ Re $s < 1$.

Proof We take the case of $M(x)$ first. Recall from 2.2.3 that, for Re $s > 1$,

$$\frac{1}{\zeta(s)} = s\int_1^\infty \frac{M(x)}{x^{s+1}}\,dx.$$

Suppose that $|M(x)| \le Cx^\alpha$ for all $x > 1$. Then, by 1.7.10, the above integral converges for all s with $\sigma > \alpha$, defining a holomorphic function $h(s)$ that agrees with $1/\zeta(s)$ for $\sigma > 1$. So $h(s)\zeta(s)$ is holomorphic and equals 1 when Re $s > 1$. By the uniqueness theorem for holomorphic functions, it follows that $h(s)\zeta(s) = 1$, and hence that $\zeta(s) \neq 0$, throughout the region Re $s > \alpha$.

Similarly, we have

$$s\int_1^\infty \frac{x - \psi(x)}{x^{s+1}}\,dx = \frac{s}{s-1} + \frac{\zeta'(s)}{\zeta(s)}$$

for Re $s > 1$ (note that this function has no pole at 1). If $x - \psi(x)$ is $O(x^\alpha)$, then the integral defines a holomorphic function $k(s)$ (without poles) for $\sigma > \alpha$. The uniqueness theorem now implies that $k(s)$ coincides with $s/(s-1) + \zeta'(s)/\zeta(s)$ on this region. Any point where $\zeta(s) = 0$ is a pole of $\zeta'(s)/\zeta(s)$, so again it follows that $\zeta(s) \neq 0$ in the region. □

So the Riemann hypothesis is *equivalent* to the statement that either $M(x)$

or $\psi(x) - x$ (and hence also $\pi(x) - \text{li}(x)$) is $O(x^\alpha)$ for all $\alpha > \frac{1}{2}$. Massive computations of actual values suggest that this is correct, both for $\pi(x) - \text{li}(x)$ and for $M(x)$. But a proof of the Riemann hypothesis remains elusive! In other words, there is a large gulf between the error estimates that have actually been proved and those suggested by calculation, and the Riemann hypothesis is essentially equivalent to the latter.

A more careful analysis shows that the Riemann hypothesis actually implies that $\psi(x) - x$ is $O[x^{1/2}(\log x)^2]$, and hence (by 5.1.7) that $\pi(x) - \text{li}(x)$ is $O(x^{1/2} \log x)$.

A rather stronger result is suspected for $M(x)$ (which, being an integer, is more pleasant to compute). In 1897, Mertens checked that $|M(x)| \leq x^{1/2}$ (with no intervening constant) for all $x \leq 10^4$ and conjectured that the same holds for all x: this statement is called the *Mertens conjecture.* Computation has now verified it at least for all $x \leq 10^{10}$, but it has been shown (with great effort) that there is an x for which $|M(x)| > 1.06\, x^{1/2}$! However, this does not disprove the possibility that the conjecture holds with an intervening constant, still less the weaker statement $M(x) = O(x^\alpha)$ for $\alpha > \frac{1}{2}$.

It is known (though we will not prove it here) that $\zeta(s)$ does have zeros on the line Re $s = \frac{1}{2}$. This sets a limit to possible error estimates: by 5.2.1, $M(x)$ and $\psi(x) - x$ are certainly not $O(x^\alpha)$ for any $\alpha < \frac{1}{2}$. It is interesting that this statement (in a slightly strengthened form) actually applies to $\psi(x) - x$ and $x - \psi(x)$ separately, not just the modulus. This part of the argument is not very hard, so we give it here.

Proposition 5.2.2 *Among the zeros of $\zeta(s)$ with* Re $s = \frac{1}{2}$, *let ρ_1 be the one with least modulus. Let $K < 1/|\rho_1|$. Then there are arbitrarily large values x_1, x_2 such that*

$$\psi(x_1) - x_1 > K x_1^{1/2}, \qquad \psi(x_2) - x_2 < -K x_2^{1/2}.$$

Proof First, assume that $\psi(x) - x$ is $O(x^\alpha)$ for all $\alpha > \frac{1}{2}$. Let $\rho_1 = \frac{1}{2} + i\gamma_1$, and suppose that $\zeta(s)$ has a zero of order m at ρ_1 (actually, it is known that $m = 1$). It follows easily that

$$\lim_{s \to \rho_1} (s - \rho_1) \frac{\zeta'(s)}{\zeta(s)} = m.$$

Choose c, and let

$$g(x) = x - \psi(x) + c x^{1/2}.$$

As in 5.2.1, our assumption implies that

$$\int_1^\infty \frac{g(x)}{x^{s+1}}\,dx = \frac{1}{s-1} + \frac{\zeta'(s)}{s\zeta(s)} + \frac{c}{s-\frac{1}{2}}$$

whenever Re $s > \frac{1}{2}$. Denote this by $f(s)$. Then

$$\lim_{s\to\rho_1}(s-\rho_1)f(s) = \frac{m}{\rho_1}. \tag{5.4}$$

Also, clearly,

$$\lim_{\sigma\to(1/2)+}(\sigma-\tfrac{1}{2})f(\sigma) = c. \tag{5.5}$$

Suppose now that, for some x_0, we have $\psi(x) - x \le cx^{1/2}$, so that $g(x) \ge 0$, for all $x \ge x_0$. Then for $s = \sigma + it$ with $\sigma > \frac{1}{2}$,

$$|f(s)| \le \int_1^{x_0} \frac{|g(x)|}{x^{\sigma+1}}\,dx + \int_{x_0}^\infty \frac{g(x)}{x^{\sigma+1}}\,dx = f(\sigma) + J, \tag{5.6}$$

where

$$J = \int_1^{x_0} \frac{|g(x)| - g(x)}{x^{\sigma+1}}\,dx.$$

Now let $s = \sigma + i\gamma_1$, so that $s - \rho_1 = \sigma - \frac{1}{2}$. By (5.4), (5.5) and (5.6),

$$\frac{1}{|\rho_1|} \le \frac{m}{|\rho_1|} = \lim_{\sigma\to(1/2)+}(\sigma-\tfrac{1}{2})|f(s)| \le \lim_{\sigma\to(1/2)+}(\sigma-\tfrac{1}{2})[f(\sigma)+J] = c.$$

Hence if $K < 1/|\rho_1|$, then K cannot be c as above, so there are arbitrarily large values of x such that $\psi(x) - x > Kx^{1/2}$.

Similarly, if $\psi(x) - x \ge cx^{1/2}$, so that $g(x) \le 0$, for all sufficiently large x, we deduce that $|f(s)| \le J - f(\sigma)$, where J is fixed, and hence that $c \le -1/|\rho_1|$.

If the above assumption is false, then there exists $\alpha > \frac{1}{2}$ such that $\psi(x) - x$ is not $O(x^\alpha)$. A somewhat different proof from the above (which we will not reproduce here) shows that one-sided versions of this statement again apply: given any K, neither $\psi(x) - x$ nor $x - \psi(x)$ is less than Kx^α for all large enough x. Of course, this implies the result stated in 5.2.2. □

For a corresponding statement for $M(x)$, see exercise 2.

A similar statement for the Liouville function $\lambda(n)$ can be obtained by quite elementary methods: see section 4.3, exercise 7.

An improvement of 5.2.2 was obtained by Littlewood in 1914: $\psi(x) - x$ is not $O(x^{1/2})$. Again one-sided versions apply: there is no constant K such that either $\psi(x) - x$ or $x - \psi(x)$ is less than $Kx^{1/2}$ for all x. The proof of Littlewood's theorem is quite hard, but it is not so hard to deduce that there is no K such that either $\pi(x) - \text{li}(x)$ or $\text{li}(x) - \pi(x)$ is less than $Kx^{1/2}/\log x$ for all x. Now $\pi(x) < \text{li}(x)$ throughout the enormous range of values that have been computed. But Littlewood's result implies that there are values of x for which $\pi(x) > \text{li}(x)$. The best current estimate for the least such x is that it is less than 10^{371}, a rather large number! This serves as a warning on the limitations of numerical evidence.

Exercises

1 Let f be a function that is continuous except at integers, and having right and left limits at integers. Let $J(x) = \int_x^\infty f(t)\,dt$. Suppose that $|J(x)| \le Kx^{-1/2}$ for all $x > 1$. Use integration by parts to show that $\int_1^\infty t^{1-s} f(t)\,dt$ is convergent when Re $s > \frac{1}{2}$. Deduce that the Riemann hypothesis would follow from the statement

$$\left| \int_x^\infty \frac{M(t)}{t^2}\,dt \right| \le \frac{K}{x^{1/2}} \qquad \text{for all } x > 1.$$

2 Let ρ_1 be a zero of the zeta function with Re $\rho_1 = \frac{1}{2}$, and let

$$K < \frac{1}{|\rho_1 \zeta'(\rho_1)|}.$$

Assuming the Riemann hypothesis, show that there are arbitrarily large values x_1, x_2 such that $M(x_1) > Kx_1^{1/2}$ and $M(x_2) < -Kx_2^{1/2}$.

3 Let $M_2(x) = \sum_{n \le x} \mu(n)^2$, and (as in 2.5.5) let $M_2(x) = x/\zeta(2) + q(x)$. Show that for a certain $c > 0$ there are arbitrarily large values of x such that $|q(x)| > cx^{1/4}$.

5.3 The zero-free region of the zeta function

Elementary results

Our objective in this section is to show that $\zeta'(s)/\zeta(s)$ and $1/\zeta(s)$ satisfy the conditions (FT3a) and (FT3b) needed for the error estimates. We can dispose of (FT3b) very quickly:

Proposition 5.3.1 *We have $\zeta(\sigma + it) \neq 0$ when $\frac{3}{4} \le \sigma \le 1$ and $|t| \le 2\frac{1}{2}$.*

Proof By 3.1.16, for $\sigma > -1$,

$$\zeta(s) = \frac{1}{s-1} + \frac{1}{2} + r_1^*(s) = \frac{s+1}{2(s-1)} + r_1^*(s),$$

where

$$|r_1^*(s)| \le \frac{|s(s+1)|}{8(\sigma+1)}.$$

So, if $\zeta(s) = 0$, then

$$\frac{1}{2|s-1|} \le \frac{|s|}{8(\sigma+1)},$$

or $4(\sigma+1) \le |s(s-1)|$. But within the region stated, $4(\sigma+1) \ge 7$, while $|s|^2 \le 7\frac{1}{4}$ and $|s-1|^2 \le 6\frac{1}{2}$, hence $|s(s-1)| < 7$. □

The rest of this section is concerned with (FT3a). The main ingredient is to establish that $\zeta(s)$ is non-zero on a region of the form $\sigma \ge 1 - c/\log t$. Although this is woefully short of the Riemann hypothesis, it still requires a good deal of work! However, a weaker version, in which $\log t$ is replaced by $(\log t)^9$, is obtained quite easily.

Either version requires the next result, which is obtained by a very straightforward modification of the proof of 3.1.6.

Proposition 5.3.2 *If $t \ge 2$ and $\sigma \ge 1 - a/\log t$ (and also $\sigma \ge \frac{1}{2}$), then*

$$|\zeta(\sigma+it)| \le e^a(\log t + 5).$$

Proof By 3.1.3, for Re $s > 0$ and any positive integer N, we have

$$\zeta(s) = \sum_{n=1}^{N} \frac{1}{n^s} - \frac{N^{1-s}}{1-s} + r_N(s),$$

where

$$|r_N(s)| \le \frac{|s|}{\sigma N^\sigma} \le \left(1 + \frac{t}{\sigma}\right)\frac{1}{N^\sigma}.$$

Since $(1-\sigma)\log t \le a$, we have $t^{1-\sigma} \le e^a$. Hence for $n \le t$ we have $n^{1-\sigma} \le e^a$, or $n^{-\sigma} \le e^a/n$. Take $N = [t]$, so that $N \le t < N+1$. The previous estimates are modified as follows:

$$\left|\sum_{n=1}^{N} \frac{1}{n^s}\right| \le e^a \sum_{n=1}^{N} \frac{1}{n} \le e^a(\log t + 1),$$

$$\left|\frac{N^{1-s}}{1-s}\right| \le \frac{N^{1-\sigma}}{t} \le \frac{e^a}{t},$$

$$|r_N(s)| \le \frac{1+2t}{N^\sigma} \le e^a \frac{1+2t}{N} \le e^a\left(2 + \frac{3}{N}\right).$$

These estimates combine to give $|\zeta(s)| \le e^a(\log t + 5)$ for $t \ge 2$. □

Note Taking $a = (1-\sigma)\log t$, we deduce that $|\zeta(\sigma+it)| \le t^{1-\sigma}(\log t + 5)$ when $\frac{1}{2} \le \sigma \le 1$ and $t \ge 2$.

By similar modification of the terms in the proof of 3.1.8, we obtain the following estimate for $|\zeta'(s)|$ (we omit the details, since this is only relevant to the weak version of (FT3a)).

Proposition 5.3.3 *If $t \ge 2$ and $\sigma \ge 1 - a/\log t$ (and also $\sigma \ge \frac{1}{2}$), then*

$$|\zeta'(\sigma+it)| \le \tfrac{1}{2}e^a(\log t + 5)^2. \qquad \square$$

Combining these estimates with 3.1.14, we now derive the weak version. If fed into the proof of the error estimates, it would have the effect that $(\log x)^{1/10}$ would appear in the conclusion instead of $(\log x)^{1/2}$.

Proposition 5.3.4 *Write $L(t) = \log t + 5$. If $t \ge 2$ and*

$$\sigma \ge 1 - \frac{1}{8L(t)^9},$$

then $\zeta(\sigma+it) \ne 0$ and in fact

$$\frac{1}{|\zeta(\sigma+it)|} \le 8L(t)^7.$$

Proof By 3.1.14, $|\zeta(1+it)| \ge 1/4L(t)^7$. Let σ be as stated, so that $1-\sigma \le 1/[8L(t)^9]$. If $\sigma \le x \le 1$, then, by 5.3.3,

$$|\zeta'(x+it)| \le \tfrac{1}{2}e^{1/8}L(t)^2 \le L(t)^2,$$

and hence

$$|\zeta(1+it) - \zeta(\sigma+it)| = \left| \int_\sigma^1 \zeta'(x+it)\,dx \right| \le (1-\sigma)L(t)^2 \le \frac{1}{8L(t)^7},$$

so that

$$|\zeta(\sigma+it)| \ge \frac{1}{8L(t)^7}. \qquad \square$$

Estimates in terms of the real part of a complex function

Write $D(z_0,R)$ for the open disc $\{z : |z-z_0| < R\}$ in the complex plane and $\overline{D}(z_0,R)$ for the closed disc $\{z : |z-z_0| \le R\}$. If a complex function f is holomorphic on $D(0,R)$, then of course it is given by a power series $\sum_{n=0}^\infty a_n z^n$ within this disc. For $r < R$, let $M(r) = \sup\{|f(z)| : |z| = r\}$. Then *Cauchy's inequality*, derived easily from the integral expression for a_n, states that $|a_n| \le M(r)/r^n$ for each n. The next result establishes a similar estimate in terms of the *real part* of $f(z)$.

Proposition 5.3.5 *Suppose that f is holomorphic on $D(0, R)$, with power series expression $f(z) = \sum_{n=0}^{\infty} a_n z^n$. Suppose also that $f(0) = 0$ (so that $a_0 = 0$). For $0 < r < R$, let*

$$A(r) = \sup\{\operatorname{Re} f(z) : |z| = r\}.$$

Then

$$|a_n| \leq \frac{2A(r)}{r^n} \qquad \text{for all } n \geq 1.$$

Proof Write $\operatorname{Re} f(z) = u(z)$ and $a_n = |a_n| e^{i\theta_n}$. Then

$$u(re^{i\theta}) = \sum_{n=1}^{\infty} |a_n| r^n \cos(n\theta + \theta_n),$$

and (for fixed r) this series converges uniformly for $-\pi \leq \theta \leq \pi$. For $n \geq 1$,

$$\int_{-\pi}^{\pi} \cos(n\theta + \theta_n)\, d\theta = 0,$$

so, by termwise integration,

$$\int_{-\pi}^{\pi} u(re^{i\theta})\, d\theta = 0. \tag{5.7}$$

Also, from the formula $2 \cos a \cos b = \cos(a + b) + \cos(a - b)$, it is clear that, for $m \neq n$,

$$\int_{-\pi}^{\pi} \cos(m\theta + \theta_m) \cos(n\theta + \theta_n) d\theta = 0,$$

so, for fixed $n \geq 1$,

$$\int_{-\pi}^{\pi} u(re^{i\theta}) \cos(n\theta + \theta_n)\, d\theta = |a_n| r^n \int_{-\pi}^{\pi} \cos^2(n\theta + \theta_n)\, d\theta = \pi |a_n| r^n. \tag{5.8}$$

So, by (5.7), (5.8) and the fact that $1 + \cos\phi \geq 0$ for any ϕ,

$$\begin{aligned} \pi |a_n| r^n &= \int_{-\pi}^{\pi} u(re^{i\theta})[1 + \cos(n\theta + \theta_n)]\, d\theta \\ &\leq A(r) \int_{-\pi}^{\pi} [1 + \cos(n\theta + \theta_n)]\, d\theta \\ &= 2\pi A(r), \end{aligned}$$

hence $|a_n| r^n \leq 2A(r)$. □

Note The factor 2 cannot be removed: see exercise 3.

Corollary 5.3.6 *Under the same conditions, we have for $|z| < r$:*

$$|f(z)| \leq 2A(r)\frac{|z|}{r - |z|},$$

$$|f'(z)| \leq 2A(r)\frac{r}{(r - |z|)^2}.$$

Note The first statement is called the *Borel-Carathéodory theorem.*

Proof By the geometric series,

$$|f(z)| \leq \sum_{n=1}^{\infty} |a_n z^n| \leq 2A(r) \sum_{n=1}^{\infty} \left|\frac{z}{r}\right|^n = 2A(r)\frac{|z|}{r - |z|}.$$

Similarly, by the series $\sum_{n=1}^{\infty} nx^{n-1} = 1/(1-x)^2$,

$$\begin{aligned} |f'(z)| \leq \sum_{n=1}^{\infty} n|a_n z^{n-1}| &\leq 2A(r) \sum_{n=1}^{\infty} \frac{n|z|^{n-1}}{r^n} \\ &= \frac{2A(r)}{r} \frac{1}{(1 - |z|/r)^2} \\ &= 2A(r)\frac{r}{(r - |z|)^2}. \end{aligned}$$

□

It is natural to apply these results to logarithms, because of the fact that $\log|f(z)| = \text{Re}\ \log f(z)$. We include the next lemma for completeness, though it is a standard result of complex analysis.

Lemma 5.3.7 *Suppose that f is holomorphic and non-zero on $D(z_0, R)$. Then a logarithm of $f(z)/f(z_0)$ is defined on this disc by*

$$h(z) = \int_{[z_0,z]} \frac{f'(\zeta)}{f(\zeta)}\, d\zeta,$$

where $[z_0, z]$ denotes the straight-line segment from z_0 to z.

Proof With this definition, we have $h'(z) = f'(z)/f(z)$, and hence

$$\frac{d}{dz}[f(z)e^{-h(z)}] = [f'(z) - h'(z)f(z)]e^{-h(z)} = 0$$

in $D(z_0, R)$. Hence $f(z)e^{-h(z)}$ is constant on the disc. Since $h(z_0) = 0$, the constant is $f(z_0)$. □

Proposition 5.3.8 *Suppose that f is holomorphic on $D(z_0, R)$. Let $r < R$. Suppose that $f(z) \neq 0$ when $|z - z_0| \leq r$ and*

$$\left|\frac{f(z)}{f(z_0)}\right| \leq e^M$$

when $|z - z_0| = r$. Then for $|z - z_0| < r$:

$$\log\left|\frac{f(z_0)}{f(z)}\right| \leq \frac{2M|z - z_0|}{r - |z - z_0|} \quad \text{and} \quad \left|\frac{f'(z)}{f(z)}\right| \leq \frac{2Mr}{(r - |z - z_0|)^2}.$$

Proof We prove the statement for the case $z_0 = 0$; the general case then follows by considering $f(z - z_0)$. Let $h(z)$ be defined as in the lemma. By our hypothesis, Re $h(z) \leq M$ when $|z| = r$. So, by 5.3.6,

$$\log\left|\frac{f(0)}{f(z)}\right| = -\text{Re } h(z) \leq |h(z)| \leq \frac{2M|z|}{r - |z|}$$

and

$$\left|\frac{f'(z)}{f(z)}\right| = |h'(z)| \leq \frac{2Mr}{(r - |z|)^2}. \qquad \square$$

Note 1 In particular, the second statement gives $|f'(z_0)/f(z_0)| \leq 2M/r$.

Note 2 If $|z - z_0| \leq \frac{1}{3}r$, then the right-hand side in the first inequality is not greater than M, so that $|f(z_0)/f(z)| \leq e^M$.

Proposition 5.3.9 *Suppose that*

(i) *f is holomorphic on $D(z_0, R)$, and $2r < R$;*

(ii) *$f(z_0) \neq 0$ and $\left|\frac{f(z)}{f(z_0)}\right| \leq e^M$ when $|z - z_0| = 2r$;*

(iii) *f has zeros at $z_1, \ldots, z_n$ (possibly repeated) in $\overline{D}(z_0, r)$.*

Then

$$\left|\frac{f'(z_0)}{f(z_0)} + \sum_{j=1}^{n} \frac{1}{z_j - z_0}\right| \leq \frac{2M}{r}.$$

Proof Again, it is enough to prove the statement for the case $z_0 = 0$. Define $g(z)$ by

$$f(z) = g(z)\prod_{j=1}^{n}(z - z_j).$$

Then g is holomorphic on $D(z_0, R)$ and non-zero when $|z| \le r$. If $|z| = 2r$, then $|z - z_j| \ge r \ge |z_j|$, so

$$\left|\frac{g(z)}{g(0)}\right| = \left|\frac{f(z)}{f(0)}\right| \prod_{j=1}^{n} \frac{|z_j|}{|z - z_j|} \le \left|\frac{f(z)}{f(0)}\right| \le e^M.$$

By the maximum modulus principle, this inequality holds for all z with $|z| \le 2r$. So, by 5.3.8 applied to the circle $|z| = r$,

$$\left|\frac{g'(0)}{g(0)}\right| \le \frac{2M}{r}.$$

Write $f_j(z) = z - z_j$, so that $f(z) = g(z)f_1(z)\dots f_n(z)$. By the product rule for differentiation,

$$\frac{f'(0)}{f(0)} = \frac{g'(0)}{g(0)} + \sum_{j=1}^{n} \frac{f_j'(0)}{f_j(0)} = \frac{g'(0)}{g(0)} - \sum_{j=1}^{n} \frac{1}{z_j}.$$

The statement follows. □

Corollary 5.3.10 *Under the conditions of 5.3.9, suppose further that* Re $z_j \le$ Re z_0 *for each* j. *Then*

$$-\text{Re}\, \frac{f'(z_0)}{f(z_0)} \le \frac{2M}{r} + \text{Re}\, \frac{1}{z_1 - z_0} \le \frac{2M}{r}.$$

Proof The extra hypothesis implies that Re $[1/(z_j - z_0)] \le 0$ for each j. The statement now follows at once from 5.3.9. □

Application to the zeta function

As with the original proof that $\zeta(1 + it) \ne 0$ in 3.1.12, we will use the simple inequality $3 + 4\cos\theta + \cos 2\theta \ge 0$, applied to Dirichlet series with non-negative coefficients. First, an elementary lemma.

Lemma 5.3.11 *For* $1 < \sigma < \frac{3}{2}$, *we have*

$$-\frac{\zeta'(\sigma)}{\zeta(\sigma)} \le \frac{1}{\sigma - 1}.$$

Proof Write $\sigma = 1 + \delta$. By the remark following 1.7.1,

$$\zeta(\sigma) \ge \frac{1}{\delta} + \frac{1}{2} = \frac{2 + \delta}{2\delta},$$

while by 1.7.12,

$$-\zeta'(\sigma) \le \frac{1}{\delta^2} + 1.$$

Since $\delta \le \frac{1}{2}$, we have $2\delta^2 \le \delta$, and hence

$$-(\sigma - 1)\frac{\zeta'(\sigma)}{\zeta(\sigma)} \le \left(\frac{1}{\delta^2} + 1\right)\frac{2\delta^2}{2+\delta} = \frac{2 + 2\delta^2}{2+\delta} \le 1. \qquad \square$$

Theorem 5.3.12 *Write $L(t) = \log t + 11$. Then $\zeta(\sigma + it) \ne 0$ when $t \ge 2$ and*

$$\sigma \ge 1 - \frac{1}{840L(t)}.$$

Proof We know from 2.1.4 that $\zeta(\sigma + it) \ne 0$ when $\sigma > 1$, so we assume that $\sigma \le 1$. Also, by 5.3.1, it is enough to prove the statement when $t \ge 2\frac{1}{2}$. Suppose, then, that $\zeta(\sigma + it_0) = 0$, where $t_0 \ge 2\frac{1}{2}$ and

$$\sigma = 1 - \frac{c'}{L},$$

in which $L = L(t_0)$ and $0 \le c' \le \frac{1}{4}$. Let

$$\sigma_0 = 1 + \frac{c}{L},$$

where c ($\le \frac{1}{2}$) is to be chosen later. Write

$$f(s) = -\frac{\zeta'(s)}{\zeta(s)}.$$

Then $f(s) = \sum_{n=1}^{\infty} \Lambda(n)/n^s$ for $\operatorname{Re} s > 1$, a Dirichlet series with non-negative coefficients. So, by 3.1.10,

$$3f(\sigma_0) + 4\operatorname{Re} f(\sigma_0 + it_0) + \operatorname{Re} f(\sigma_0 + 2it_0) \ge 0. \tag{5.9}$$

By lemma 5.3.11, $f(\sigma_0) \le L/c$.

We will use 5.3.10 to estimate $\operatorname{Re} f(\sigma_0 + it_0)$ and $\operatorname{Re} f(\sigma_0 + 2it_0)$. Write $s_0 = \sigma_0 + it_0$. First, by 2.2.4 and 1.7.1,

$$\frac{1}{|\zeta(s_0)|} \le \zeta(\sigma_0) \le 1 + \frac{L}{c}.$$

Let $r = (\log L/)L$. Note that since $L > 10$, we have $r < \frac{1}{10}\log 10 < \frac{1}{4}$. Let $z = x + iy$ be a point on the circle $|z - s_0| = 2r$. We apply 5.3.2 to estimate $|\zeta(z)|$. Now $y > t_0 - \frac{1}{2} \ge 2$; also $y < t_0 + \frac{1}{2}$, hence certainly $\log y + 5 < L$. Further,

$$x > 1 - 2r = 1 - \frac{2\log L}{L} > 1 - \frac{2\log L}{\log y},$$

so, by 5.3.2,

$$|\zeta(z)| \le e^{2\log L}(\log y + 5) \le L^2 L = L^3.$$

We will ensure that c is chosen so that

$$L^3 \ge 1 + \frac{L}{c}.$$

Hence we have

$$\left|\frac{\zeta(z)}{\zeta(s_0)}\right| \le \left(1 + \frac{L}{c}\right) L^3 \le L^6.$$

Now $\zeta(s)$ has the assumed zero at the point $s_1 = \sigma + it_0$ within the disc $|s - s_0| < r$, and $s_1 - s_0 = -(c + c')/L$. Also, of course, there are no zeros having Re $s \ge \sigma_0$. So, by 5.3.10, we have

$$\text{Re } f(\sigma_0 + it_0) \le \frac{12\log L}{r} - \frac{L}{c + c'} = 12L - \frac{L}{c + c'}. \tag{5.10}$$

The same argument applies to the point $\sigma_0 + 2it_0$. The condition $\log y + 5 < L$ is still satisfied, and there may or may not be any zeros within the disc of radius r. So we have

$$\text{Re } f(\sigma_0 + 2it_0) \le 12L. \tag{5.11}$$

Hence, by (5.9), (5.10) and (5.11):

$$\frac{3L}{c} + 48L - \frac{4L}{c + c'} + 12L \ge 0,$$

so that

$$\frac{4}{c + c'} \le \frac{3}{c} + 60.$$

Now choose $c = \frac{1}{120}$. Then $4/(c + c') \le 420$, so

$$c' \ge \frac{1}{105} - \frac{1}{120} = \frac{1}{840}.$$

Since $L \ge 11$, the required condition $L^3 \ge L/c + 1 = 120L + 1$ is satisfied. □

Note 1 We have not tried very hard to find the best possible constants, but it is easy to see that the constants will improve if we restrict to larger values of t.

Note 2 With a different constant again, we can restate the theorem with $\log t$ replacing $L(t)$, as in the original formulation of condition (FT3a).

Finally, we use 5.3.8 again to give bounds for $1/\zeta(s)$ and $\zeta'(s)/\zeta(s)$ within the zero-free region.

Theorem 5.3.13 *Write $L(t) = \log t + 11$ and $c = \frac{1}{840}$. There are constants A, B, k such that, whenever $s = \sigma + it$ with $t \geq 2 + c$ and $\sigma \geq 1 - c/6L(t)$, we have*

$$\frac{1}{|\zeta(s)|} \leq AL(t)^3,$$

$$\left|\frac{\zeta'(s)}{\zeta(s)}\right| \leq BL(t)\left(\log L(t) + k\right).$$

We can take $A = 56/c^2$, $B = 9/c$ and $k = \frac{1}{2}\log(8/c)$.

Proof If $\sigma > 1 + c/6L(t)$, then $1/|\zeta(s)| \leq \zeta(\sigma) \leq 1 + 6L(t)/c$, and 5.3.11 shows that a similar estimate holds for $\zeta'(s)/\zeta(s)$. So we assume that $\sigma \leq 1 + c/6L(t)$. Fix $t \geq 2 + c$. Write $L(t) = L$ and $c_0 = c/6$. Let $r = c/L$ and

$$s_0 = 1 + \frac{c_0}{L} + it.$$

Then

$$\frac{1}{|\zeta(s_0)|} \leq 1 + \frac{L}{c_0} = 1 + \frac{6L}{c} < \frac{7L}{c}.$$

Consider $z = x + iy$ such that $|z - s_0| \leq r$. Then $2 \leq t - c \leq y \leq t + c$, which certainly implies that $L(y) \leq \frac{6}{5}L$. Meanwhile,

$$x \geq 1 - \frac{5c}{6L} \geq 1 - \frac{c}{L(y)},$$

so $\zeta(z) \neq 0$. Also, $\log y + 5 < L$, so by 5.3.2 we have $|\zeta(z)| \leq e^c L$, and hence

$$\left|\frac{\zeta(z)}{\zeta(s_0)}\right| \leq \frac{7}{c}e^c L^2 \leq \frac{8}{c}L^2,$$

(since $e^c < \frac{8}{7}$). This is e^M in the notation of 5.3.8. Now suppose that $|\sigma - 1| \leq c_0/L$ and let $s = \sigma + it$. Then $|s - s_0| \leq 2c_0/L = \frac{1}{3}r$, so by 5.3.8 (in the form given by the subsequent note 2),

$$\frac{1}{|\zeta(s)|} \leq \frac{e^M}{|\zeta(s_0)|} \leq \frac{56}{c^2}L^3.$$

Also, by the second statement in 5.3.8,

$$\left|\frac{\zeta'(s)}{\zeta(s)}\right| \leq \frac{2Mr}{\frac{4}{9}r^2} = \frac{9M}{2r} = \frac{9L}{2c}(2\log L + 2k) = \frac{9L}{c}(\log L + k),$$

where $2k = \log(8/c)$. □

Note 1 The bound for $\zeta'(s)/\zeta(s)$ could be replaced by $B'L(t)^2$ or $B''(\log t)^2$, with new constants B', B''.

Note 2 The vigilant reader will notice that this theorem has only been stated for $t \geq 2+c$. Of course, similar inequalities (with suitable constants) apply for $2 \leq t \leq 2+c$ simply because $\zeta(s) \neq 0$ in this region.

The method can be adapted without serious change to Dirichlet L-functions. This leads to error estimates in the theorems of section 4.4 of the same form as those obtained for $\psi(x)$ and $M(x)$. However, the constants appearing will depend on the modulus k.

With slightly more effort, one can amend both estimates in 5.3.13 to ones of the form $A \log t$; however, this makes no serious difference to the error estimates.

With a lot more effort, it has been shown that the zero-free region includes at least

$$\sigma \geq 1 - C(\log t)^{-2/3}(\log\log t)^{-1/3}.$$

This leads to the following slightly improved error estimate in the prime number theorem:

$$|\psi(x) - x| \leq Kx \exp[-c(\log x)^{3/5}(\log\log x)^{-1/5}].$$

See, for example, [Ivić].

Exercises

1. Let f be holomorphic on $D(0,R)$, with $f(0) = 0$, and let $2r < R$. With the notation of the text, show that $M(r) \leq 2A(2r)$.

2. Let f be holomorphic on $D(0,R)$. For $r < R$, let $A(r)$ be defined as in the text. By applying the maximum modulus principle to $e^{f(z)}$, prove that Re $f(z) \leq A(r)$ when $|z| < r$, with equality only ocurring if f is constant. Now prove the Borel-Carathéodory theorem by applying Schwarz's lemma to $f(z)/[2A(r) - f(z)]$.

3. Let $f(z) = z/(1+z)$. Use the geometric series to write down the power series for $f(z)$ when $|z| < 1$. Show that Re $f(z) \leq \frac{1}{2}$ for such z (so that the factor 2 cannot be removed in 5.3.5).

4. Formulate and prove an estimate analogous to 5.3.2 for the Dirichlet L-function corresponding to a non-principal character.

6
An "elementary" proof of the prime number theorem

Both proofs of the prime number theorem given in chapter 3 are "analytic" in the sense that they rely fundamentally on theorems of complex analysis, following the design set out by Riemann. As we have seen, these methods ultimately prove the theorem very satisfactorily, exhibiting it as one case of a wider family of theorems. However, one would surely expect to be able to prove a theorem about integers without resorting to concepts like holomorphic complex functions, which are seemingly far removed from pure number theory. Such a proof, reverting to the legacy of Chebyshev, would be called "elementary". Ideas from real analysis, such as integral estimation for series, would be allowed (the theorem is, after all, a statement about a limit), but not those from complex analysis. Defined in this way, the term "elementary" is not to be confused with "simple"!

This question attracted a great deal of interest after the original proof had become known. G. H. Hardy, a leading mathematician of the time, predicted in 1921 that such a proof was "extraordinarily unlikely", but that if one were to be found, it would involve such basic new ideas that it would "cause the whole theory to be rewritten".

An elementary proof was finally discovered by A. Selberg and P. Erdös in 1948. They worked in cooperation, but eventually published different versions separately, both based on a formula discovered by Selberg. The history of their on-off collaboration is somewhat tangled (for one account, see [Nath]). Many alternative versions and refinements of the proof have appeared since then. Nearly all make use of Selberg's formula in one way or another. For a survey of these various methods, see [Dia]. A method of Daboussi that is not based on Selberg's formula is described in [T&MF]. Even after 50 years of tidying up, one would not describe any of these methods as simple. Furthermore, they have not really transformed the theory in the way that Hardy predicted. They do not present the prime number theorem as part of a wider scheme of theorems, and have not, as yet, been developed to give error estimates as strong as that of de la Vallée Poussin. So it may be said that both parts of Hardy's prediction have proved to be wrong!

In this chapter, we give an account of one of these versions. With minor simplifications, it follows a method set out by N. Levinson in 1969 [Lev]. The work can be divided into two parts, with quite distinct characters, which we present in two separate sections. The goal is to prove the prime number theorem in the form $\psi(x)/x \to 1$. The main prerequisites, all to be found in our chapters 1 and 2, are Chebyshev's upper estimate, Möbius inversion, the identity $\Lambda * u = \ell$ and the boundedness of $\sum_{n\leq x} \Lambda(n)/n - \log x$.

6.1 Framework of the proof

Define

$$R(x) = \begin{cases} 0 & \text{if } x < 1, \\ \psi(x) - x & \text{if } x \geq 1. \end{cases}$$

Note that $R(x)$ assumes both positive and negative values. Our objective is to show that $\psi(x)/x \to 1$ or, equivalently, that $R(x)/x \to 0$, as $x \to \infty$.

We assume Chebyshev's upper estimate in the form $\psi(x) \leq 2x$ for all $x \geq 1$ (see 2.4.6). We also recall from 2.6.1 and 2.6.2 the equivalent statements:

$$\sum_{n\leq x} \frac{\Lambda(n)}{n} = \log x + O(1), \qquad \int_1^x \frac{\psi(t)}{t^2}\, dt = \log x + O(1).$$

These facts translate into the following properties of $R(x)$.

Proposition 6.1.1 *We have:*

(i) $|R(x)| \leq x$ *for all* $x > 0$;

(ii) $\displaystyle\int_1^x \frac{R(t)}{t^2}\, dt$ *is bounded for all* $x \geq 1$.

Proof Statement (i) is immediate from Chebyshev's estimate, and statement (ii) restates 2.6.2, since

$$\int_1^x \frac{R(t)}{t^2}\, dt = \int_1^x \frac{\psi(t)}{t^2}\, dt - \log x. \qquad \square$$

Of course, $R(x)$ is discontinuous at primes (and powers of primes). A "smoothed-out" alternative is given by integration: define (for $x > 0$)

$$S(x) = \int_0^x \frac{R(t)}{t}\, dt = \int_1^x \frac{\psi(t)}{t}\, dt - (x-1).$$

The properties above translate as follows.

Proposition 6.1.2 *We have:*

(i) $|S(x_2) - S(x_1)| \le x_2 - x_1$ *whenever* $x_2 > x_1 > 0$ *(i.e., S is Lipschitz); in particular,* $|S(x)| \le x$ *and S is continuous;*

(ii) $\displaystyle\int_1^x \frac{S(t)}{t^2}\, dt$ *is bounded for all* $x \ge 1$.

Proof Statement (i) follows from the fact that $|R(t)/t| \le 1$ for all t. To prove (ii), we reverse the order of repeated integrals to obtain

$$\begin{aligned}\int_1^x \frac{S(t)}{t^2}\, dt &= \int_1^x \frac{1}{t^2} \int_1^t \frac{R(u)}{u}\, du\, dt \\ &= \int_1^x \frac{R(u)}{u} \int_u^x \frac{1}{t^2}\, dt\, du \\ &= \int_1^x \frac{R(u)}{u} \left(\frac{1}{u} - \frac{1}{x}\right) du \\ &= \int_1^x \frac{R(u)}{u^2}\, du - \frac{S(x)}{x}.\end{aligned}$$

This is bounded, by the two statements in 6.1.1. (There need be no worry about reversing an integral containing the discontinuous function $R(u)$, since one can imagine the exercise performed separately on each interval of the form $[n, n+1)$.) □

The following Tauberian theorem (a result of the same sort as 3.4.1) ensures that it will be enough to prove our statement for $S(x)$ instead of $R(x)$.

Proposition 6.1.3 *Suppose that $A(x)$ is non-negative and increasing for $x \ge 1$, and that*

$$\frac{1}{x} \int_1^x \frac{A(t)}{t}\, dt \to 1 \quad \text{as } x \to \infty.$$

Then

$$\frac{A(x)}{x} \to 1 \quad \text{as } x \to \infty.$$

Proof Write

$$F(x) = \int_1^x \frac{A(t)}{t}\, dt.$$

Note that $F(\lambda x)/x \to \lambda$ as $x \to \infty$. Take $\varepsilon > 0$ and $x \ge 1$. For t between x

and $(1+\varepsilon)x$, we have

$$\frac{A(t)}{t} \geq \frac{A(x)}{(1+\varepsilon)x},$$

hence

$$F[(1+\varepsilon)x] - F(x) = \int_x^{(1+\varepsilon)x} \frac{A(t)}{t}\,dt \geq \frac{\varepsilon}{1+\varepsilon}A(x). \tag{6.1}$$

But

$$\frac{1}{x}\{F[(1+\varepsilon)x] - F(x)\} \to \varepsilon \quad \text{as } x \to \infty,$$

so for all sufficiently large x we have

$$F[(1+\varepsilon)x] - F(x) \leq \varepsilon(1+\varepsilon)x. \tag{6.2}$$

By (6.1) and (6.2), for such x we have $A(x) \leq (1+\varepsilon)^2 x$.

Similarly,

$$F(x) - F[(1-\varepsilon)x] \leq \frac{\varepsilon}{1-\varepsilon}A(x),$$

and hence for sufficiently large x we have $A(x) \geq (1-\varepsilon)^2 x$. □

Applying this result with $A(x) = \psi(x)$, we see that if we can show that $S(x)/x \to 0$ as $x \to \infty$, then it will follow that $\psi(x)/x \to 1$, as required.

We now make the further substitution

$$W(x) = \frac{S(e^x)}{e^x}.$$

Clearly, $|W(x)| \leq 1$ for all $x > 0$, and our objective is to show that $W(x) \to 0$ as $x \to \infty$. The properties of S translate as follows:

Proposition 6.1.4 *We have*

(i) $|W(x_2) - W(x_1)| \leq 2(x_2 - x_1)$ *whenever* $x_2 > x_1 > 0$;

(ii) $\int_0^x W(t)\,dt$ *is bounded for all* $x > 0$.

Proof (i) Write $y_j = e^{x_j}$ for $j = 1, 2$. By 6.1.2(i), we have

$$\begin{aligned} |y_1 S(y_2) - y_2 S(y_1)| &= |y_1[S(y_2) - S(y_1)] + (y_1 - y_2)S(y_1)| \\ &\leq (y_1 + |S(y_1)|)(y_2 - y_1) \\ &\leq 2y_1(y_2 - y_1), \end{aligned}$$

so

$$\left|\frac{S(y_2)}{y_2} - \frac{S(y_1)}{y_1}\right| \leq 2\left(1 - \frac{y_1}{y_2}\right),$$

or in other words,

$$|W(x_2) - W(x_1)| \leq 2(1 - e^{-(x_2 - x_1)}).$$

The elementary inequality $e^{-x} \geq 1 - x$ now gives $|W(x_2) - W(x_1)| \leq 2(x_2 - x_1)$.

(ii) The substitution $t = \log u$ gives

$$\int_0^x W(t)\,dt = \int_0^x e^{-t} S(e^t)\,dt = \int_1^{e^t} \frac{S(u)}{u^2}\,du,$$

which is bounded, by 6.1.2(ii). □

Upper limits. Let f be any bounded real-valued function on (a, ∞) for some a. Define $s_f(x) = \sup_{y \geq x} f(y)$. Then $s_f(x)$ is clearly a decreasing function of x. It is also bounded below, so it tends to a limit as $x \to \infty$. This limit is called the *upper limit* of $f(x)$ as $x \to \infty$, denoted by $\limsup_{x\to\infty} f(x)$.

Of course, if $f(x)$ tends to a limit (say α) as $x \to \infty$, then $\limsup_{x\to\infty} f(x) = \alpha$. However, every bounded function has an upper limit, while many of them do not tend to any limit. Beyond this, all we really need to know about upper limits is summarized in the following two points:

(i) If $\limsup_{x\to\infty} f(x) = \alpha$, and $\varepsilon > 0$ is given, then it follows at once from the definition that there exists x_1 such that $f(x) \leq \alpha + \varepsilon$ for all $x \geq x_1$. Conversely, this statement implies that $\limsup_{x\to\infty} f(x) \leq \alpha$.

(ii) If $\limsup_{x\to\infty} |f(x)| = 0$, then $\lim_{x\to\infty} f(x) = 0$.

Let f be bounded on $(0, \infty)$, and let

$$g(x) = \frac{1}{x} \int_0^x f(t)\,dt.$$

It is an easy exercise (which the reader may care to attempt) to show that $\limsup_{x\to\infty} g(x) \leq \limsup_{x\to\infty} f(x)$. Inequality can easily occur here: for example, if $f(x) = \cos x$, then $\limsup_{x\to\infty} f(x) = 1$, while $g(x) = \sin x / x$, which tends to 0 as $x \to \infty$. The next result, a crucial step in the elementary proof of the prime number theorem, shows that the function $|W(x)|$ behaves in this way unless its limit is 0.

Proposition 6.1.5 *Let $W(x)$ be defined as above, and let*

$$\alpha = \limsup_{x\to\infty} |W(x)|,$$

$$\beta = \limsup_{x\to\infty} \frac{1}{x} \int_0^x |W(t)|\,dt.$$

Then either $\alpha = 0$ or $\beta < \alpha$.

Proof By 6.1.4, there exists M such that $|\int_{x_1}^{x_2} W(x)\, dx| \le M$ for all x_1, x_2 (note that the statement to be proved refers to the integral of $|W(x)|$, not $W(x)$!). Assume that $\alpha > 0$, and choose $a > \alpha$ (but not greater than 2α). Then there exists x_0 such that $|W(x)| \le a$ for all $x \ge x_0$. Now let $x_1 \ge x_0$, and consider the integral of $|W(x)|$ on the interval $I = [x_1, x_1 + h]$, where h ($\ge 2\alpha$) is to be chosen. If either $W(x) \ge 0$ or $W(x) \le 0$ on I, then clearly

$$\int_{x_1}^{x_1+h} |W(x)|\, dx \le M.$$

Otherwise, since $W(x)$ is continuous, it is zero at some point z of I. Then, by 6.1.4(i), we have $|W(x)| = |W(x) - W(z)| \le 2|x - z|$ for all x. In particular, if $|x - z| \le a/2$, this is better than the estimate $|W(x)| \le a$ that applies generally on I. Since $h \ge a$, at least one of $z \pm a/2$ (say $z + a/2$) is in I, and we have

$$\int_{z}^{z+a/2} |W(x)|\, dx \le \int_{z}^{z+a/2} 2(x - z)\, dx = \frac{a^2}{4}.$$

Combining this with the estimate $|W(x)| \le a$ on the rest of I, we obtain

$$\int_{x_1}^{x_1+h} |W(x)|\, dx \le \left(h - \frac{a}{2}\right) a + \frac{a^2}{4} = a\left(h - \frac{a}{4}\right) < a\left(h - \frac{\alpha}{4}\right).$$

To ensure that this inequality also holds in the case where $W(x)$ has one sign, we choose h so that $M \le \alpha(h - \alpha/4)$, in other words, $h \ge M/\alpha + \alpha/4$. So we take h to be the greater of this quantity and α (note: we are taking care to ensure that h does not vary with a). Write $h - \alpha/4 = \rho h$, where (clearly) $\rho < 1$. In either case, we have

$$\int_{x_1}^{x_1+h} |W(x)|\, dx \le \rho a h.$$

Now take any $x \ge x_1 + h$, and let n be the integer such that $x_0 + nh \le x < x_0 + (n + 1)h$. Write $\int_0^{x_0} |W| = C$. Then

$$\int_0^x |W(x)|\, dx \le C + (n + 1)\rho a h,$$

so

$$\frac{1}{x}\int_0^x |W(x)|\, dx \le \frac{C}{x} + \left(1 + \frac{1}{n}\right)\rho a.$$

Now $C/x + \rho a/n \to 0$ as $x \to \infty$, and hence we have $\beta \le \rho a$. This is true for all $a > \alpha$, so in fact $\beta \le \rho\alpha < \alpha$. □

To complete the proof of the prime number theorem, we will show that in fact $\beta = \alpha$. This depends on work of a completely different nature, which we present in the next section.

Exercises

1 Prove that

$$\limsup_{x\to\infty}[f(x) + g(x)] \le \limsup_{x\to\infty} f(x) + \limsup_{x\to\infty} g(x),$$

and give an example of inequality.

2 Let f be a bounded function, and let

$$g(x) = \frac{1}{x}\int_0^x f(t)\, dt.$$

Show that $\limsup_{x\to\infty} g(x) \le \limsup_{x\to\infty} f(x)$.

3 Show that the result of 6.1.3 also holds when the limit is 0.

6.2 Selberg's formulae and completion of the proof

Our goal in this section is to prove, in the notation of 6.1.5, that $\beta = \alpha$. First we shall establish a formula of Selberg involving $R(x)$ (or $S(x)$) that is of some interest in itself. At this stage it is far from transparent how Selberg's formula is related to the stated objective. However, we go on to show that it leads to a bound for *values* of $S(x)$ in terms of *integrals* involving the same function. This in turn leads to the required statement, with $W(x)$ only making its appearance at the last moment.

Our overall objective is, of course, to show that $R(x)$ $(= \psi(x) - x)$ is ultimately small compared to x. We start from the idea of trying to use Möbius inversion to find a bound for $R(x)$. Recall 2.2.8: if F is a function on $[1, \infty)$ and $G(x) = \sum_{n\le x} F(x/n)$, then

$$F(x) = \sum_{n\le x} \mu(n) G\left(\frac{x}{n}\right). \tag{6.3}$$

Hence a bound for $|G(x)|$ will lead to one for $|F(x)|$. Roughly, the idea is to apply this with $F(x) = R(x)$. As the next lemma shows, one can do better by taking $R(x) + c$ for a suitably chosen c.

Lemma 6.2.1 *Let* $F(x) = R(x) + \gamma + 1$ *and* $G(x) = \sum_{n\le x} F(x/n)$. *Then* $|G(x)| \le \log x + 2$ *for all* $x \ge 1$.

Proof Since $\Lambda * u = \ell$, we have by 1.8.5

$$\sum_{n \le x} \log n = \sum_{n \le x} (\Lambda * u)(n) = \sum_{n \le x} \psi\left(\frac{x}{n}\right),$$

while by 1.4.3 we have

$$\sum_{n \le x} \log n = x \log x - x + 1 + r_1(x),$$

where $|r_1(x)| \le \log x$ for all $x \ge 1$.

At the same time, by 1.4.11,

$$\sum_{n \le x} \frac{x}{n} = x \sum_{n \le x} \frac{1}{n} = x \log x + \gamma x + r_2(x),$$

where $|r_2(x)| \le 1$. By including the term $\gamma+1$ in $F(x)$, we essentially achieve cancellation of the x term as well as of $x \log x$. More exactly, since $x = [x] + \{x\}$, we have

$$\begin{aligned}\sum_{n \le x} F\left(\frac{x}{n}\right) &= \sum_{n \le x} \psi\left(\frac{x}{n}\right) - \sum_{n \le x} \frac{x}{n} + (\gamma + 1)[x] \\ &= -(\gamma + 1)\{x\} + r_1(x) + 1 - r_2(x).\end{aligned}$$

Clearly, this is between $\log x + 2$ and $-(\log x + \gamma + 1)$. □

Note It would be a bit too sloppy just to say that $G(x) = O(\log x)$, since this is not correct near 1 (note that $G(1) = F(1) = \gamma$), and we shall be considering sums involving $G(x/n)$ with roughly half the values x/n between 1 and 2.

Let us see what happens if we now apply (6.3). By Abel summation in the form 1.3.6(ii),

$$\sum_{n \le x} \log \frac{x}{n} = \sum_{n \le x} (\log x - \log n) = \int_1^x \frac{[t]}{t}\, dt < x,$$

and hence

$$|R(x) + \gamma + 1| \le \sum_{n \le x} \left| G\left(\frac{x}{n}\right) \right| \le \sum_{n \le x} \left(\log \frac{x}{n} + 2 \right) \le 3x.$$

So we have dismally failed to show that $R(x)$ is eventually small compared to x! What we have actually achieved is a new proof of Chebyshev's upper estimate, though not with a very good estimate of the constant.

Selberg's formulae will be derived by using a variant of (6.3) with a logarithmic factor introduced, as follows.

Lemma 6.2.2 *For all $n \geq 1$, we have* $\Lambda(n) = -\sum_{k|n} \mu(k) \log k$.

Proof In convolution notation, this says $\Lambda = -(\ell\mu) * u$. Now $\ell(\mu * u) = \ell e_1 = 0$, since $\ell(1) = \log 1 = 0$ and $e_1(n) = 0$ for $n \geq 2$. So by 1.8.7,

$$\begin{aligned} 0 = \ell(\mu * u) &= (\ell\mu) * u + \mu * (\ell u) \\ &= (\ell\mu) * u + \mu * \ell \\ &= (\ell\mu) * u + \Lambda. \end{aligned}$$

□

Note The corresponding identity for Dirichlet series is $\zeta'/\zeta = -\zeta(1/\zeta)'$.

Proposition 6.2.3 (the Tatuzawa-Iseki identity) *Let F be a function on $[1, \infty)$ and let $G(x) = \sum_{n \leq x} F(x/n)$. Then for $x \geq 1$,*

$$\sum_{k \leq x} \mu(k) \log \frac{x}{k} \, G\Big(\frac{x}{k}\Big) = F(x) \log x + \sum_{n \leq x} \Lambda(n) F\Big(\frac{x}{n}\Big).$$

Proof First, by (6.3),

$$F(x) \log x = \log x \sum_{k \leq x} \mu(k) G\Big(\frac{x}{k}\Big). \tag{6.4}$$

Meanwhile, we also have

$$\begin{aligned} \sum_{k \leq x} \mu(k) \log k \, G\Big(\frac{x}{k}\Big) &= \sum_{k \leq x} \mu(k) \log k \sum_{j \leq x/k} F\Big(\frac{x}{jk}\Big) \\ &= \sum_{jk \leq x} F\Big(\frac{x}{jk}\Big) \mu(k) \log k \\ &= \sum_{n \leq x} F\Big(\frac{x}{n}\Big) \sum_{k|n} \mu(k) \log k \\ &= -\sum_{n \leq x} F\Big(\frac{x}{n}\Big) \Lambda(n), \qquad \text{by 6.2.2.} \end{aligned}$$

Combining this equality with (6.4), we obtain the statement. □

We now establish two equivalent versions of Selberg's formula.

Proposition 6.2.4 *Let $R(x) = \psi(x) - x$. Then, for all $x \geq 1$,*

$$R(x) \log x + \sum_{n \leq x} \Lambda(n) R\Big(\frac{x}{n}\Big) = O(x), \tag{6.5}$$

$$\psi(x) \log x + \sum_{n \leq x} \Lambda(n) \psi\Big(\frac{x}{n}\Big) = 2x \log x + O(x). \tag{6.6}$$

Proof First, we show that (6.5) and (6.6) are equivalent. Recall from 2.6.1 that $\sum_{n\leq x}\Lambda(n)/n = \log x + O(1)$. Hence

$$\begin{aligned}\sum_{n\leq x}\Lambda(n)R\left(\frac{x}{n}\right) &= \sum_{n\leq x}\Lambda(n)\psi\left(\frac{x}{n}\right) - \sum_{n\leq x}\Lambda(n)\frac{x}{n}\\ &= \sum_{n\leq x}\Lambda(n)\psi\left(\frac{x}{n}\right) - x\log x + O(x),\end{aligned}$$

so the left-hand side of (6.5) equates to

$$\psi(x)\log x + \sum_{n\leq x}\Lambda(n)\psi\left(\frac{x}{n}\right) - 2x\log x + O(x).$$

This establishes the stated equivalence. We now prove (6.5). Write $c = \gamma+1$, and let $J(x)$ be the quantity considered in 6.2.3, with $F(x) = R(x)+c$. From the right-hand side, we have

$$\begin{aligned}J(x) &= F(x)\log x + \sum_{n\leq x}\Lambda(n)F\left(\frac{x}{n}\right)\\ &= R(x)\log x + c\log x + \sum_{n\leq x}\Lambda(n)R\left(\frac{x}{n}\right) + c\psi(x)\\ &= R(x)\log x + \sum_{n\leq x}\Lambda(n)R\left(\frac{x}{n}\right) + O(x),\end{aligned}$$

where we have used Chebyshev's estimate $\psi(x) = O(x)$ in the last line. Meanwhile, 6.2.1 tells us that $|G(x)| \leq \log x + 2$ for all x, so, from the left-hand side in 6.2.3, we have

$$|J(x)| \leq \sum_{k\leq x}\log\frac{x}{k}\left(\log\frac{x}{k}+2\right).$$

Now there is a constant C such that

$$\log x(\log x + 2) \leq Cx^{1/2}$$

for all $x \geq 1$ (this is elementary; the reader can fill in the details). Hence

$$\begin{aligned}|J(x)| &\leq C\sum_{k\leq x}\left(\frac{x}{k}\right)^{1/2}\\ &= Cx^{1/2}\sum_{k\leq x}\frac{1}{k^{1/2}}\\ &\leq Cx^{1/2}.2x^{1/2}\\ &= 2Cx.\end{aligned}$$

□

Let us pause to reflect on this result, say in the form (6.5). It tells us that the given expression is "small" in the sense that we only know, at this stage, that each term separately is $O(x \log x)$. Of course, the term in which we are really interested is $R(x) \log x$. Indeed, if we knew that this term on its own were $O(x)$, then we would already have the desired result, in the pleasant form $\psi(x) - x = O(x/\log x)$. The effect of using 6.2.3 instead of ordinary Möbius inversion is to introduce the second term in (6.5). The remaining task, in a sense, is to disentangle the two terms. This will eventually be achieved along the lines we have indicated, but the end result will not guarantee anything as neat as each term in (6.5) being $O(x)$.

Another equivalent version of Selberg's formula is given in the next result.

Corollary 6.2.5 *Let $q(n) = \Lambda(n) \log n + (\Lambda * \Lambda)(n)$. Then:*

$$\sum_{n \leq x} q(n) = 2x \log x + O(x),$$

$$\sum_{n \leq x} (q(n) - 2 \log n) = O(x).$$

Proof By Abel's summation formula,

$$\sum_{n \leq x} \Lambda(n) \log n = \psi(x) \log x - \int_1^x \frac{\psi(t)}{t} \, dt = \psi(x) \log x + O(x),$$

since $\psi(t) \leq 2t$ for all t. Also, by 1.8.4,

$$\sum_{n \leq x} (\Lambda * \Lambda)(n) = \sum_{n \leq x} \Lambda(n) \psi\left(\frac{x}{n}\right).$$

So the first statement is equivalent to (6.6). The second statement follows, since $\sum_{n \leq x} \log n = x \log x + O(x)$. □

Note The notation Λ_2 is often used instead of q.

Next, we show that R can be replaced by S in (6.5).

Proposition 6.2.6 *For all $x \geq 1$, we have*

$$S(x) \log x + \sum_{n \leq x} \Lambda(n) S\left(\frac{x}{n}\right) = O(x).$$

Proof Let $1 \leq t \leq x$. Since $R(u) = 0$ for $u < 1$, we have

$$\sum_{n \leq x} \Lambda(n) R\left(\frac{t}{n}\right) = \sum_{n \leq t} \Lambda(n) R\left(\frac{t}{n}\right).$$

Hence, dividing through by t in (6.5), we have

$$R(t)\frac{\log t}{t} + \frac{1}{t}\sum_{n\leq x}\Lambda(n)R\left(\frac{t}{n}\right) = O(1).$$

This means that the left-hand side is bounded by some constant, independent of t and x. So we can integrate to obtain

$$\int_1^x R(t)\frac{\log t}{t}\,dt + \sum_{n\leq x}\Lambda(n)\int_1^x R\left(\frac{t}{n}\right)\frac{1}{t}\,dt = O(x).$$

We consider these two terms separately. The substitution $t = nu$ gives

$$\int_1^x R\left(\frac{t}{n}\right)\frac{1}{t}\,dt = \int_{1/n}^{x/n}\frac{R(u)}{u}\,du = \int_1^{x/n}\frac{R(u)}{u}\,du = S\left(\frac{x}{n}\right).$$

Also,

$$\begin{aligned}\int_1^x \frac{S(t)}{t}\,dt &= \int_1^x \frac{1}{t}\int_1^t \frac{R(u)}{u}\,du\,dt\\ &= \int_1^x \frac{R(u)}{u}\int_u^x \frac{1}{t}\,dt\,du\\ &= \int_1^x \frac{R(u)}{u}(\log x - \log u)\,du\\ &= S(x)\log x - \int_1^x R(u)\frac{\log u}{u}\,du.\end{aligned}$$

But this quantity is $O(x)$, since $|S(t)/t| \leq 1$ for all t. Hence

$$\int_1^x R(u)\frac{\log u}{u}\,du = S(x)\log x + O(x),$$

and the statement follows. □

Now we use the second term to give a bound for the first term, but in a way that introduces $q(n)$ (for which we have a good estimate of the partial sums) instead of $\Lambda(n)$.

Proposition 6.2.7 *There is a constant K_1 such that, for all $x \geq 1$,*

$$|S(x)|(\log x)^2 \leq \sum_{n\leq x} q(n)\left|S\left(\frac{x}{n}\right)\right| + K_1 x\log x.$$

Proof Recall from 2.6.1 that $\sum_{k\leq x}\Lambda(k)/k \leq \log x + 3$, which in turn is not greater than $\rho\log x$ for all $x \geq 2$ (for a suitable ρ). Since $\Lambda(1) = 0$, this bound actually holds for all $x \geq 1$. By 6.2.6, there is a constant C such that, for each $k \leq x$,

$$\left| S\left(\frac{x}{k}\right) \log \frac{x}{k} + \sum_{j \leq x/k} \Lambda(j) S\left(\frac{x}{jk}\right) \right| \leq C\frac{x}{k}.$$

Multiply by $\Lambda(k)$ and add for $k \leq x$: we obtain

$$|A + B| \leq Cx \sum_{k \leq x} \frac{\Lambda(k)}{k} \leq \rho C x \log x,$$

where

$$A = \sum_{k \leq x} \Lambda(k) S\left(\frac{x}{k}\right) \log \frac{x}{k},$$

$$B = \sum_{k \leq x} \Lambda(k) \sum_{j \leq x/k} \Lambda(j) S\left(\frac{x}{jk}\right).$$

Writing $jk = n$, we have

$$B = \sum_{n \leq x} S\left(\frac{x}{n}\right) \sum_{jk=n} \Lambda(j)\Lambda(k) = \sum_{n \leq x} (\Lambda * \Lambda)(n) S\left(\frac{x}{n}\right).$$

Hence

$$A + B = \log x \sum_{k \leq x} \Lambda(k) S\left(\frac{x}{k}\right) + \sum_{n \leq x} S\left(\frac{x}{n}\right) q_1(n),$$

where

$$q_1(n) = (\Lambda * \Lambda)(n) - \Lambda(n) \log n.$$

Of course, $|q_1(n)| \leq q(n)$. By 6.2.6 again,

$$\sum_{k \leq x} \Lambda(k) S\left(\frac{x}{k}\right) = -S(x) \log x + O(x).$$

It follows that

$$-S(x)(\log x)^2 + \sum_{n \leq x} S\left(\frac{x}{n}\right) q_1(n) = O(x \log x),$$

so, for some constant K_1, we have

$$\begin{aligned} |S(x)|(\log x)^2 &\leq \left| \sum_{n \leq x} S\left(\frac{x}{n}\right) q_1(n) \right| + K_1 x \log x \\ &\leq \sum_{n \leq x} q(n) \left| S\left(\frac{x}{n}\right) \right| + K_1 x \log x. \end{aligned}$$

□

By 6.2.5, $q(n)$ behaves roughly like $2\log n$ in sums, and we now show that it can indeed be replaced by $2\log n$ in 6.2.7. This is a distinct second use of Selberg's formulae.

Proposition 6.2.8 *We have*

$$\sum_{n\le x} q(n)\left|S\left(\frac{x}{n}\right)\right| = 2\sum_{n\le x}\log n\left|S\left(\frac{x}{n}\right)\right| + O(x\log x).$$

Proof Write $a(n) = q(n) - 2\log n$ and $A(n) = \sum_{j\le n} a(j)$. By 6.2.5, there exists C such that $|A(n)| \le Cn$ for all n. Since $S(x/m) = 0$ for $m > x$, Abel summation gives

$$\sum_{n\le x} a(n)\left|S\left(\frac{x}{n}\right)\right| = \sum_{n\le x} A(n)\left[\left|S\left(\frac{x}{n}\right)\right| - \left|S\left(\frac{x}{n+1}\right)\right|\right].$$

Since $|S(x_2) - S(x_1)| \le 2(x_2 - x_1)$ for $x_2 > x_1$, the modulus of this sum is not greater than

$$2C\sum_{n\le x} n\left(\frac{x}{n} - \frac{x}{n+1}\right) = 2Cx\sum_{n\le x}\frac{1}{n+1} \le 2Cx(\log x + 1). \qquad \square$$

Next, we replace the discrete sum in 6.2.8 by an integral.

Proposition 6.2.9 *There is a constant C' such that, for all $x \ge 1$,*

$$\sum_{n\le x}\log n\left|S\left(\frac{x}{n}\right)\right| \le \int_1^x \log t\left|S\left(\frac{x}{t}\right)\right|\,dt + C'x.$$

Proof Let $n \le x$ and take t such that $n \le t < n+1$. Again using the fact that $|S(x_2) - S(x_1)| \le x_2 - x_1$, we have

$$\left|S\left(\frac{x}{n}\right)\right| \le \left|S\left(\frac{x}{t}\right)\right| + \frac{x}{n} - \frac{x}{t} \le \left|S\left(\frac{x}{t}\right)\right| + x\left(\frac{1}{n} - \frac{1}{n+1}\right),$$

so that

$$\log n\left|S\left(\frac{x}{n}\right)\right| \le \int_n^{n+1}\log t\left|S\left(\frac{x}{t}\right)\right|\,dt + x\frac{\log n}{n(n+1)}.$$

The series

$$\sum_{n=1}^{\infty}\frac{\log n}{n(n+1)}$$

is convergent, say to A (where in fact $A < 1$). Adding the inequalities above (and remembering that $S(x/t) = 0$ when $t > x$), we obtain

$$\sum_{n \le x} \log n \left| S\left(\frac{x}{n}\right)\right| \le \int_1^x \log t \left| S\left(\frac{x}{t}\right)\right| \, dt + Ax. \qquad \square$$

Combining the last three results, we obtain:

Proposition 6.2.10 *There is a constant K_2 such that, for all $x \ge 1$,*

$$|S(x)|(\log x)^2 \le 2\int_1^x \log t \left| S\left(\frac{x}{t}\right)\right| \, dt + K_2 x \log x.$$

If $W(y) = e^{-y}S(e^y)$, then

$$|W(y)| \le \frac{2}{y^2}\int_0^y (y-u)|W(u)| \, du + \frac{K_2}{y}.$$

Proof The first statement has just been proved. With the substitutions $x = e^y$ and $t = e^u$, it becomes

$$y^2|S(e^y)| \le 2\int_0^y |S(e^{y-u})|ue^u \, du + K_2 y e^y.$$

With the further substitution $S(e^y) = e^y W(y)$, we obtain

$$y^2 e^y |W(y)| \le 2\int_0^y e^y |W(y-u)|u \, du + K_2 y e^y,$$

or

$$|W(y)| \le \frac{2}{y^2}\int_0^y |W(y-u)|u \, du + \frac{K_2}{y}.$$

Finally, note that

$$\int_0^y |W(y-u)|u \, du = \int_0^y (y-v)|W(v)| \, dv. \qquad \square$$

Note that this result gives a bound for $|W(y)|$ in terms of an integral involving W. The required statement about upper limits now follows with no trouble.

Proposition 6.2.11 *Let α, β be defined as in 6.1.5. Then $\alpha \le \beta$.*

Proof Note first that

$$\int_0^x (x-u)|W(u)| \, du = \int_0^x |W(u)| \int_u^x 1 \, dv \, du = \int_0^x \int_0^v |W(u)| \, du \, dv.$$

Choose $\varepsilon > 0$. By the definition of β, there exists x_1 such that, for all $v \geq x_1$,

$$\frac{1}{v}\int_0^v |W(u)|\,du \leq \beta + \varepsilon.$$

Hence for $x > x_1$,

$$\int_{x_1}^x \int_0^v |W(u)|\,du\,dv \leq \int_{x_1}^x (\beta+\varepsilon)v\,dv = \tfrac{1}{2}(\beta+\varepsilon)(x^2 - x_1^2).$$

Also, since $|W(u)| \leq 1$ for all u,

$$\int_0^{x_1}\int_0^v |W(u)|\,du\,dv \leq \int_0^{x_1} v\,dv = \tfrac{1}{2}x_1^2.$$

Hence

$$\int_0^x (x-u)|W(u)|\,du \leq \tfrac{1}{2}(\beta+\varepsilon)x^2 + \tfrac{1}{2}x_1^2,$$

so, by 6.2.10,

$$|W(x)| \leq \beta + \varepsilon + \frac{x_1^2}{x^2} + \frac{K_2}{x}.$$

It follows that $\alpha \leq \beta + \varepsilon$. This holds for any $\varepsilon > 0$, so in fact $\alpha \leq \beta$. □

Together with 6.1.5, this shows that $\alpha = 0$, or in other words, $W(x) \to 0$ as $x \to \infty$. So we have completed the elementary proof of the prime number theorem in the form:

Theorem 6.2.12 *For Chebyshev's function ψ, we have*

$$\frac{\psi(x)}{x} \to 1 \quad \text{as } x \to \infty.$$ □

We conclude with some remarks on variations of this proof. First, it is possible to dispense with the move from $R(x)$ to $S(x)$, and hence with the steps connected with this move. One then uses $V(x) = e^{-x}R(e^x)$ instead of $W(x)$. The cost is that R and V do not have the Lipschitz property of S and W. This is quite easily overcome in the estimations 6.2.8 and 6.2.9 involving S (see the exercises), but the trivial estimation for W in 6.1.5 has to be replaced by a more difficult estimation of the increase in $\psi(x)/x$, requiring another application of Selberg's formulae (see [H&Wr]).

Second, an alternative to 6.2.7 (again working with R rather than S) is to establish that

$$|R(x)|\log x \leq \sum_{n \leq x}\left|R\left(\frac{x}{n}\right)\right| + O(x\log\log x). \tag{6.7}$$

Note that this is rather like (6.5) with the $\Lambda(n)$ replaced by 1. The steps

6.2.8, 6.2.10 and 6.2.11 then disappear or become trivial. However, the proof of (6.7) is long and intricate (see [Nath]).

Third, what about the companion result $M(x)/x \to 0$ as $x \to \infty$? In fact, this statement and the prime number theorem can each be deduced from the other by reasonably short elementary methods (see [Ap], chapter 4). However, it makes sense to ask how the method of this chapter can be adapted to give a direct proof. The function $R(x)$ is now $M(x)$ itself. The equivalent of (6.5) is much easier: the identity $\Lambda * \mu = -\ell\mu$ gives immediately

$$\sum_{n \le x} \Lambda(n) M\left(\frac{x}{n}\right) = -\sum_{n \le x} \mu(n) \log n = -M(x) \log x + O(x).$$

However, the step corresponding to 6.2.7 introduces $q(n)$, so that Selberg's formulae are still needed. The overall plan is the same as before, but it becomes easier to dispense with the move to our $S(x)$: the difficulty mentioned above is overcome by the fact that $M(x)$ takes only integer values and satisfies $|M(q) - M(p)| \le |q - p|$ for integers p, q.

Exercises

1 Show that $q = \mu * \ell^2$.

2 Show that $|R(y) - R(x)| \le F(y) - F(x)$ for $y > x$, where $F(x) = \psi(x) + x$. Now use Abel summation and Chebyshev's estimate to show that

$$\sum_{n \le x} \log n \left[F\left(\frac{x}{n}\right) - F\left(\frac{x}{n+1}\right) \right] \le 3x.$$

(This shows how to adapt 6.2.9 for R instead of S.)

3 *(Avoiding the move to W in 6.2.11).* Let

$$\alpha = \limsup_{x \to \infty} \frac{|S(x)|}{x}, \qquad \beta = \limsup_{x \to \infty} \frac{1}{\log x} \int_1^x \frac{|S(t)|}{t^2}\, dt.$$

Using the property of S given in 6.2.10, deduce that $\alpha \le \beta$.

4 Let $q_1 = \Lambda\ell$ and $q_2 = \Lambda * \Lambda$. Show that

$$Q_1(x) = \sum_{p \in P[x]} (\log p)^2 + O(x^{1/2} \log x),$$

$$Q_2(x) = \sum \{\log p \log q : p, q \text{ prime},\ pq \le x\} + O(x).$$

Appendix A
Complex functions of a real variable

A complex-valued function of a real variable can be expressed in the form $f(x) = u(x) + iv(x)$, where $u(x)$, $v(x)$ are real-valued functions.

Limits and derivatives of such functions are defined formally in exactly the same way as for functions from $\mathbb{R}$ to $\mathbb{R}$. Hence, for example,

$$f'(x_0) = \lim_{x \to x_0} \frac{f(x) - f(x_0)}{x - x_0}$$

when this limit exists. It is trivial to check that the statement $f(x) \to a + ib$ as $x \to x_0$ is equivalent to $u(x) \to a$ and $v(x) \to b$ as $x \to x_0$, and that $f'(x) = u'(x) + iv'(x)$ when $u'(x)$ and $v'(x)$ exist. The rules for limits and derivatives of sums and products are the same as for functions from $\mathbb{R}$ to $\mathbb{R}$, and are proved in exactly the same way. The same applies to the chain rule, in the following sense: *If $f : \mathbb{R} \to \mathbb{C}$ and $g : \mathbb{C} \to \mathbb{C}$ are differentiable, then so is $g \circ f$, and $(g \circ f)'(x) = g'[f(x)]f'(x)$.*

Integration of a function $f = u + iv$ as above is defined in the obvious way:

$$\int_a^b (u + iv) = \int_a^b u + i \int_a^b v.$$

Integrals of derivatives work as usual: $\int_a^b f' = f(b) - f(a)$, since the corresponding statements hold for u and v.

As for real-valued functions, $\int_a^\infty f(x)\,dx$ means $\lim_{R \to \infty} \int_a^R f(x)\,dx$ if this limit exists. Clearly, this equates to $\int_a^\infty u(x)\,dx + i \int_a^\infty v(x)\,dx$. Note that $|u(x)| \le |f(x)|$, and similarly for $|v(x)|$. So by the integral comparison test for real functions, convergence of $\int_a^\infty |f(x)|\,dx$ implies convergence of $\int_a^\infty u(x)\,dx$ and $\int_a^\infty v(x)\,dx$, and hence of $\int_a^\infty f(x)\,dx$.

Finally, we show that the standard inequality for the integral of $|f(x)|$ still applies. This requires a proof somewhat different from the real case.

Proposition A.1 *Let f be a continuous, complex-valued function on a real interval $[a, b]$. Then*

$$\left| \int_a^b f(x)\, dx \right| \le \int_a^b |f(x)|\, dx.$$

Proof Let $\int_a^b f = I$. Assume that $I \neq 0$, and let $\alpha = \overline{I}/|I|$. Then $|\alpha| = 1$ and $\int_a^b \alpha f = \alpha I = |I|$. Now write $\alpha f(x) = u_1(x) + i v_1(x)$. Since $\int_a^b \alpha f$ is real, it equals $\int_a^b u_1$. But certainly $u_1(x) \le |\alpha f(x)| = |f(x)|$ for all x, so

$$|I| = \int_a^b u_1(x)\, dx \le \int_a^b |f(x)|\, dx. \qquad \square$$

Appendix B

Double series and multiplication of series

Suppose that numbers $a_{j,k}$ (real or complex) are given for each pair (j,k) of positive integers. One can think of these numbers as forming an infinite matrix. We say that the "repeated sum"

$$\sum_{j=1}^{\infty}\sum_{k=1}^{\infty} a_{j,k}$$

converges to S if, firstly, $\sum_{k=1}^{\infty} a_{j,k}$ converges (say to a_{j*}) for each j and, secondly, $\sum_{j=1}^{\infty} a_{j*}$ converges to S. Note that a_{j*} is the sum of row j of the matrix, so this repeated sum (if it exists) is the sum of the row sums. Similarly, the other repeated sum $\sum_{k=1}^{\infty}\sum_{j=1}^{\infty} a_{j,k}$ (if it exists) is the sum of the column sums.

In general, even if both repeated sums exist, they can be unequal. For example, let

$$a_{j,j} = 1, \quad a_{j,j+1} = -1, \quad a_{j,k} = 0 \quad \text{for other } j,k.$$

Then the two repeated sums are 1 and 0 (this is easily seen on writing out the matrix).

There are also many ways in which the terms $a_{j,k}$ can be rearranged to form a single series $\sum_{n=1}^{\infty} c_n$. More exactly, this means that, for some bijection σ of $\mathbb{N} \times \mathbb{N}$ onto $\mathbb{N}$, we have $c_{\sigma(j,k)} = a_{j,k}$. A typical example is to define σ by enumerating the pairs (j,k) along successive diagonals. Note that repeated sums are not a case of this kind of rearrangement.

We shall use the following elementary fact repeatedly: if $a_j \geq 0$ for all j and $\sum_{j=1}^{\infty} a_j = A$, then $\sum_{j=1}^{J} a_j \leq A$ for all J.

Theorem B.1 *Suppose that the repeated sum* $\sum_{j=1}^{\infty}\sum_{k=1}^{\infty} |a_{j,k}|$ *converges.*

Then the repeated sums

$$\sum_{j=1}^{\infty}\sum_{k=1}^{\infty} a_{j,k}, \qquad \sum_{k=1}^{\infty}\sum_{j=1}^{\infty} a_{j,k}$$

both converge to the same sum, say S. Further, if $\sum_{n=1}^{\infty} c_n$ is any series obtained by arranging the terms $a_{j,k}$ as a single series, then $\sum_{n=1}^{\infty} c_n$ converges to S.

Proof We shall prove the statements for the case where the terms $a_{j,k}$ are all non-negative. The statements for any *real* $a_{j,k}$ then follow in the usual way by expressing $a_{j,k}$ as $a_{j,k}^{+} - a_{j,k}^{-}$, where

$$a^{+} = \begin{cases} a & \text{if } a \geq 0, \\ 0 & \text{if } a < 0, \end{cases} \qquad a^{-} = \begin{cases} 0 & \text{if } a \geq 0, \\ -a & \text{if } a < 0. \end{cases}$$

The statements for *complex* $a_{j,k}$ follow by considering the real and imaginary parts separately.

We suppose, then, that $a_{j,k} \geq 0$ for all j, k and that the first repeated sum equals S. Let $\sum_{k=1}^{\infty} a_{j,k} = a_{j*}$. Fix J, K. By the remark above, $\sum_{j=1}^{J} a_{j*} \leq S$. Also, we have $\sum_{k=1}^{K} a_{j,k} \leq a_{j*}$ for each j, so

$$\sum_{j=1}^{J}\sum_{k=1}^{K} a_{j,k} \leq S. \tag{B.1}$$

Now choose $\varepsilon > 0$. There exists j_0 such that $\sum_{j=1}^{j_0} a_{j*} \geq S - \varepsilon$. Also, there exists k_0 such that, for each $j \leq j_0$, we have $\sum_{k=1}^{k_0} a_{j,k} \geq a_{j*} - \varepsilon/j_0$, so that

$$\sum_{j=1}^{j_0}\sum_{k=1}^{k_0} a_{j,k} \geq S - 2\varepsilon \tag{B.2}$$

(and similarly if j_0, k_0 are replaced by larger numbers).

For a fixed k, we have $0 \leq a_{j,k} \leq a_{j*}$ for all j, since $a_{j,k}$ is one term of the series defining a_{j*}. So, by the comparison test for series, $\sum_{j=1}^{\infty} a_{j,k}$ is convergent, say to a_{*k}. Letting $J \to \infty$ in (B.1), we see that $\sum_{k=1}^{K} a_{*k} \leq S$ for all K. Meanwhile, it follows from (B.2) that for all $K \geq k_0$, we have $\sum_{k=1}^{K} a_{*k} \geq S - 2\varepsilon$. Hence $\sum_{k=1}^{\infty} a_{*k} = S$.

Now let $\sum_{n=1}^{\infty} c_n$ be a rearrangement of the terms $a_{j,k}$ as a single series. For any N, it is clear from (B.1) that $\sum_{n=1}^{N} c_n \leq S$. On the other hand, if N is so large that $\{c_1, c_2, \ldots, c_N\}$ contains all the terms $a_j b_k$ with $j \leq j_0$, $k \leq k_0$, then (B.2) shows that $\sum_{n=1}^{N} c_n \geq S - 2\varepsilon$. So $\sum_{n=1}^{\infty} c_n = S$. □

Corollary B.2 *Suppose that the series $\sum_{j=1}^{\infty} a_j$ and $\sum_{k=1}^{\infty} b_k$ (of real or complex terms) are absolutely convergent, with sums A and B, respectively. Let $\sum_{n=1}^{\infty} c_n$ be any series obtained by arranging the terms $a_j b_k$ as a single series. Then $\sum_{n=1}^{\infty} c_n$ is absolutely convergent, with sum AB.*

Proof Clearly,

$$\sum_{j=1}^{\infty}\sum_{k=1}^{\infty} a_j b_k = \sum_{j=1}^{\infty} a_j B = AB,$$

and similarly for the other repeated sum (so equality of the repeated sums is trivial in this case). The same applies to the terms $|a_j b_k|$, so the statement is a case of theorem B.1. □

As with any series, one can now group the terms c_n in brackets (as, for example, in the way dictated by a power series or a Dirichlet series). The resulting series is still convergent to the same sum.

Actually, for Dirichlet products (and hence for the product of two Dirichlet series as in 1.8.1), absolute convergence of *one* series is sufficient, as the following shows (this is another result due to Mertens). We switch notation to $a(n)$ for consistency with the main text.

Theorem B.3 *Suppose that $\sum_{n=1}^{\infty} a(n) = A$, $\sum_{n=1}^{\infty} b(n) = B$, and $\sum_{n=1}^{\infty} |a(n)|$ is convergent (say to A^*). Let c be the Dirichlet convolution $a * b$. Then $\sum_{n=1}^{\infty} c(n)$ converges to AB.*

Proof Write $A(x) = \sum_{n \le x} a(n)$, similarly $B(x)$. By 1.8.4,

$$C(n) = \sum_{j=1}^{n} a(j)B(n/j) = \sum_{j=1}^{n} a(j)[B(n/j) - B] + A(n)B.$$

Write this as $D(n) + A(n)B$. Now $A(n)B \to AB$ as $n \to \infty$, so the statement follows if we can show that $D(n) \to 0$. Let M be such that $|B(x) - B| \le M$ for all x. Take $\varepsilon > 0$. There exist n_0, n_1 such that $\sum_{j>n_0} |a(j)| \le \varepsilon$ and $|B(x) - B| \le \varepsilon$ for all $x \ge n_1$. Take $n > n_0 n_1$. Then

$$\sum_{j \le n_0} |a(j)[B(n/j) - B]| \le \varepsilon \sum_{j \le n_0} |a(j)| \le A^* \varepsilon,$$

$$\sum_{j > n_0} |a(j)[B(n/j) - B]| \le M \sum_{j > n_0} |a(j)| \le M\varepsilon,$$

hence $D(n) \le (A^* + M)\varepsilon$. So $D(n) \to 0$, as required. □

Appendix C
Infinite products

Let (u_n) be a sequence of *non-zero* numbers, real or complex. Let $p_n = u_1u_2 \dots u_n$ for each n. If (p_n) converges to a *non-zero* limit as $n \to \infty$, we say that the infinite product $\prod_{n=1}^{\infty} u_n$ *converges to* p. Otherwise, including the case when $p_n \to 0$, we say that the product is *divergent*.

Example Let

$$u_n = 1 - \frac{1}{(n+1)^2} = \frac{n(n+2)}{(n+1)^2}.$$

After cancellation, one sees that

$$p_n = \frac{n+2}{2(n+1)},$$

which tends to 2 as $n \to \infty$, so that $\prod_{n=1} u_n = \frac{1}{2}$.

Proposition C.1 *If* $\prod_{n=1}^{\infty} u_n$ *converges to* p *and* $\prod_{n=1}^{\infty} v_n$ *converges to* q, *then* $\prod_{n=1}^{\infty} u_n v_n$ *converges to* pq *and* $\prod_{n=1}^{\infty} (1/p_n)$ *converges to* $1/p$.

Proof Let $u_1u_2 \dots u_n = p_n$ and $v_1v_2 \dots v_n = q_n$, so that $p_n \to p$ and $q_n \to q$ as $n \to \infty$. Then

$$\prod_{j=1}^{n} u_j v_j = p_n q_n \to pq \quad \text{as } n \to \infty,$$

$$\prod_{j=1}^{n} \frac{1}{u_j} = \frac{1}{p_n} \to \frac{1}{p} \quad \text{as } n \to \infty.$$

□

Proposition C.2 *If* $a_n \neq -1$ *for all* n *and* $\sum_{n=1}^{\infty} |a_n|$ *is convergent, then* $\prod_{n=1}^{\infty} (1 + a_n)$ *is convergent.*

Proof Let

$$p_n = (1 + a_1)(1 + a_2) \dots (1 + a_n)$$

for each n. Then, for $n \geq 2$, we have $p_n = (1+a_n)p_{n-1}$, so that $p_n - p_{n-1} = a_n p_{n-1}$ (*).

Suppose that $\sum_{n=1}^{\infty} |a_n| = S$. Now

$$|1 + a_n| \leq 1 + |a_n| \leq e^{|a_n|}$$

for each n, and hence $|p_n| \leq e^S$ for all n. Therefore $|a_n p_{n-1}| \leq e^S |a_n|$, so $\sum_{n=2}^{\infty} |a_n p_{n-1}|$ is convergent, by the comparison test for series. By (*), this implies that $\sum_{n=2}^{\infty}(p_n - p_{n-1})$ is convergent. This means that

$$\sum_{r=2}^{n}(p_r - p_{r-1}) = p_n - p_1$$

tends to a (finite) limit as $n \to \infty$. Hence p_n tends to a limit (say p) as $n \to \infty$.

We still have to show that $p \neq 0$. To do this, we will show that the product $\prod_{n=1}^{\infty}(1 + a_n)^{-1}$ also converges to a finite limit, say q. Then $pq = 1$, so both p and q are non-zero. Now

$$\frac{1}{1 + a_n} = 1 - b_n,$$

where

$$b_n = \frac{a_n}{1 + a_n}.$$

Now $1 + a_n \to 1$ as $n \to \infty$, so for sufficiently large n we have $|1 + a_n| > \frac{1}{2}$, hence $|b_n| \leq 2|a_n|$. Therefore $\sum_{n=1}^{\infty} |b_n|$ is convergent, so, by what we have already proved, $\prod_{n=1}^{\infty}(1 - b_n)$ does indeed converge to a finite limit. □

Appendix D
Differentiation under the integral sign

We will not prove this result in its most general form, but rather in the form needed for our applications to Dirichlet integrals and series.

Lemma D.1 *For complex s with $|s| \leq \frac{1}{2}$, we have*

$$|e^s - 1 - s| \leq |s|^2.$$

Proof By the series for e^s,

$$\begin{aligned}
|e^s - 1 - s| &\leq \frac{1}{2!}|s|^2 + \frac{1}{3!}|s|^3 + \cdots \\
&\leq \frac{1}{2}\left(|s|^2 + |s|^3 + \cdots\right) \\
&= \frac{|s|^2}{2(1-|s|)} \\
&\leq |s|^2,
\end{aligned}$$

since $2(1-|s|) \geq 1$. □

Proposition D.2 *Let $b > a > 0$. Suppose that the function f is bounded and integrable (in the sense of Riemann) on $[a, b]$. For complex s, let*

$$I(s) = \int_a^b \frac{f(x)}{x^s}\, dx.$$

Then $I(s)$ is holomorphic for all s, with

$$I'(s) = -\int_a^b \frac{f(x)\log x}{x^s}\, dx.$$

Proof Suppose that $|f(x)| \leq M$ on $[a, b]$. Fix s_0 and let

$$J = \int_a^b \frac{f(x) \log x}{x^{s_0}}\, dx.$$

In accordance with the definition of differentiation, we have to show that

$$\frac{I(s) - I(s_0)}{s - s_0} \to -J \qquad \text{as } s \to s_0. \tag{D.1}$$

Now

$$I(s) - I(s_0) + (s - s_0)J = \int_a^b \left(\frac{f(x)}{x^s} - \frac{f(x)}{x^{s_0}} + (s - s_0) \frac{f(x) \log x}{x^{s_0}} \right) dx. \tag{D.2}$$

We estimate this integral, using the fact that $|\int_a^b g(x)\, dx| \leq \int_a^b |g(x)|\, dx$ (see appendix A). By the lemma, if s is complex, c is real and $|cs| \leq \frac{1}{2}$, then

$$|e^{cs} - 1 - cs| \leq |cs|^2.$$

Let $x > 0$ and take $c = -\log x$: if $|s \log x| \leq \frac{1}{2}$, this gives

$$\left| \frac{1}{x^s} - 1 + s \log x \right| \leq |s \log x|^2.$$

Substituting $s - s_0$ for s and dividing both sides by $|x^{s_0}| = x^{\sigma_0}$, we obtain

$$\left| \frac{1}{x^s} - \frac{1}{x^{s_0}} + \frac{(s - s_0) \log x}{x^{s_0}} \right| \leq |s - s_0|^2 \frac{(\log x)^2}{x^{\sigma_0}},$$

provided that $|(s - s_0) \log x| \leq \frac{1}{2}$. Hence the integrand in (D.2) satisfies the same estimate, multiplied by M. Let K be such that $|\log x| \leq K$ for $a \leq x \leq b$, and let $K|s - s_0| \leq \frac{1}{2}$. Then

$$|I(s) - I(s_0) + (s - s_0)J| \leq |s - s_0|^2 A,$$

where

$$A = \int_a^b \frac{(\log x)^2}{x^{\sigma_0}}\, dx.$$

In other words,

$$\left| \frac{I(s) - I(s_0)}{s - s_0} + J \right| \leq A|s - s_0|.$$

Clearly, this implies (D.1). □

Appendix E
The O, o notation

Suppose that $f(x)$, $h(x)$ are defined for all x greater than some value x_0, with $h(x) > 0$ on this range. We write

$$f(x) = O[h(x)] \qquad \text{for } x > x_1$$

if there is a constant C such that $|f(x)| \leq Ch(x)$ for all $x > x_1$. The words "for $x > x_1$" may be left out if this is the whole range of values where $f(x)$ is defined. We also write "$f(x) = O[h(x)]$ as $x \to \infty$" if there is *some* x_1 such that the above occurs. The notation applies equally whether $h(x)$ is large, small or constant. Usually, $h(x)$ will be a simple function, like a power of x. The statement $f(x) = O(1)$ just means that $f(x)$ is bounded on the range indicated. So, for example, we have

$$\sin x = O(1), \qquad x^2 \sin x = O(x^2), \qquad \frac{\sin x}{x} = O(1/x)$$

for $x > 1$.

We list some obvious facts:

(i) *If $f(x)$ and $g(x)$ are $O[h(x)]$ for $x > x_1$, then so is $f(x) + g(x)$.*

(ii) *If $f(x) = O[g(x)]$ and $g(x) = O[h(x)]$, then $f(x) = O[h(x)]$.*

(iii) *If $f(x) = O[h(x)]$ and $g(x) > 0$, then $f(x)g(x) = O[h(x)g(x)]$.*

By a natural extension of this notation, we write

$$f(x) = g(x) + O[h(x)] \qquad \text{for } x > x_1$$

if $f(x) = g(x) + q(x)$, where $q(x) = O[h(x)]$. In the same way, one might write

$$f(x) = g(x) + O[h_1(x)] + O[h_2(x)],$$

with the obvious meaning. In this situation, if we also know that $h_2(x) =$

$O[h_1(x)]$, then it is clear that the second O term can be absorbed into the first one, giving simply $f(x) = g(x) + O[h_1(x)]$. This is a typical step in calculations of this sort.

There are obvious further variations of the notation. For instance, x could be replaced by an integer variable n. Or we may wish to consider x tending to 0 (or 0^+) instead of infinity: the stated inequality must then hold for all x close enough to 0. So, for example, $\sinh x = x + O(x^3)$ as $x \to 0$.

The O notation is an economical way of stating relationships of the type described, but it makes no distinction between the cases when the constant C is $\frac{1}{4}$ or 25,000, and when used in the $x \to \infty$ form, it gives no indication of how large is "large enough". Rather limited use is made of it in this book: when the implied constants are simple and can be estimated easily, statements are given in a form that incorporates them.

A second notational device is as follows: one writes

$$f(x) = o[h(x)] \quad \text{as } x \to \infty$$

if $f(x)/h(x) \to 0$ as $x \to \infty$. Hence, for example, $(\log x)^4 = o(x)$ as $x \to \infty$. Obvious equivalents of the previous remarks apply. Note that if $f(x)$ is $O(x^\alpha)$ (as $x \to \infty$) for all $\alpha > 0$, then it is actually $o(x^\alpha)$ for all $\alpha > 0$ (since it is $O(x^{\alpha/2})$). Again, the notation is used in statements of the form

$$f(x) = g(x) + o[h(x)] \quad \text{as } x \to \infty.$$

Appendix F
Computing values of $\pi(x)$

We shall describe the *principles* of various methods for computing $\pi(N)$ for a given N. We leave it to readers to translate these principles into their chosen programming language or package.

The simplest method is the well-known "sieve of Eratosthenes". It operates as follows to create a list of primes up to a chosen N, from which one can read off values of $\pi(n)$. Start with the full list of integers from 2 to N. Delete all multiples of 2, apart from 2 itself. Then delete all multiples of 3 (apart from 3 itself). Since 4 has already been deleted, move on to 5 and delete its multiples. Continue the process, deleting multiples of each $m \leq N^{1/2}$ unless m has already been deleted. Note that when deleting multiples of m, it is enough to start at m^2, since the numbers km (for $2 \leq k < m$) have already been deleted. When we have finished, the remaining numbers are the primes in $[2, N]$, since every composite number in this interval is a multiple of some $m \leq N^{1/2}$, and consequently has been deleted.

To implement this process on a computer, what we really do is construct the function

$$f(n) = \begin{cases} 1 & \text{if } n \text{ is prime,} \\ 0 & \text{otherwise} \end{cases}$$

(in the notation of this book, f is u_P). We start by setting $f(n) = 1$ for $2 \leq n \leq N$. The act of "deleting" an integer n equates to changing $f(n)$ from 1 to 0. The programming steps are essentially as follows (in which $N_1 = [N^{1/2}]$):

set $f(n) = 1$ for $2 \leq n \leq N$
for $2 \leq m \leq N_1$, increasing in steps of 1
if $f(m) = 1$ (that is, if m not already "deleted")
for $m^2 \leq n \leq N$, increasing in steps of m, set $f(n) = 0$.

This produces our function $f(n)$. To *count* the prime numbers, define the function `pi` (the usual π) by:

$$\texttt{pi}(1) = 0$$
$$\texttt{pi}(n) = \texttt{pi}(n-1) + f(n).$$

It is wasteful to use $f(n)$ as a *list* of prime numbers, since it includes all the non-primes. For this purpose, we define the function `pr` such that $\texttt{pr}(i)$ is the ith prime:

$i = 1$
$n = 2$
if $f(n) = 1$, set $\texttt{pr}(i) = n$ and increase i by 1
increase n by 1
repeat if $n < N$.

Some modern computers with sufficient memory will handle this for quite large values of N (possibly up to $N = 10^6$). However, storage problems will eventually be encountered. To overcome this, divide the long interval $[2, N]$ into blocks of length k, where k is roughly $N^{1/2}$. First, apply the sieve to find and store the primes up to $N^{1/2}$. Then, for each block separately, use these known primes to sieve the block, leaving only the primes. This is done using the same memory space each time: a new function indexed by $1, 2, \ldots, k$ overrides the previous one. All we need to store is the *number* of primes in each block. Of course, for blocks beyond the first one, we must identify the first multiple of m occurring in the block. Suppose that the block consists of the integers from $nk+1$ to $(n+1)k$. If nk has remainder r when divided by m, then the required multiple is $nk + m - r$ (note that nk is the last member of the *previous* block).

The Meissel-Lehmer method. Even with the modification just described, the sieve of Eratosthenes is expensive in terms of storage. To push the limits further, what is needed is a method that calculates $\pi(x)$ without calculating particular primes. Such a method was originated by D. E. F. Meissel in 1870, who performed the remarkable feat of calculating $\pi(10^9)$ nearly a century before the age of computers (unfortunately, a minor error resulted in Meissel's value being wrong by 56). His method was developed further by D. H. Lehmer in 1959. It reduces the computation of $\pi(x)$ to the evaluation of $\pi(n)$ for $n \leq x^{2/3}$ together with a number of shorter sieving exercises. An outline follows.

Let (p_n) be the sequence of prime numbers (in order). For $n \geq 1$, define

$S(x, n) =$ the set of integers $m \leq x$ that are *not* divisible by $p_1, \ldots, p_n$;
$\phi(x, n) =$ the number of members of $S(x, n)$.

Also, define $S(x,0)$ to be the set of all positive integers $m \le x$. Clearly, $S(x,n)$ is the result of sieving out multiples of $p_1,\ldots,p_n$ (this time including these numbers themselves). It contains 1 and those numbers ($\le x$) that are products of primes greater than p_n.

The number $\phi(x,n)$ is easily evaluated for the first few values of n. Indeed, $S(x,1)$ is just the set of odd integers not greater than x, and $S(x,2)$ is the set of integers of the form $6r+1$ or $6r+5$. Similarly, $S(x,3)$ only retains 8 numbers in each block of 30. However, $S(x,5)$ is already described in terms of blocks of length $2.3.5.7.11 = 2310$, and this line of thought is not much use for n any larger.

At the opposite extreme, if $n \ge \pi(x)$, then $p_{n+1} > x$, so that $\phi(x,n) = 1$. Further:

Lemma F.1 *If $\pi(x^{1/2}) \le n \le \pi(x)$ (equivalently, $p_{n+1} > x^{1/2}$ and $p_n \le x$), then $\phi(x,n) = \pi(x) - n + 1$.*

Proof The stated equivalence is elementary. Under these conditions, the members of $S(x,n)$ are 1 and the primes p such that $p_n < p \le x$, since any product of two such primes is greater than x. □

When $n = \pi(x^{1/2})$, this lemma essentially restates the sieve of Eratosthenes. We shall relate $\pi(x)$ to $\phi(x,n)$ for a smaller n, using the following recurrence relation.

Lemma F.2 *For all $n \ge 1$, we have*

$$\phi(x,n) = \phi(x,n-1) - \phi\left(\frac{x}{p_n}, n-1\right).$$

Proof Members of $S(x,n-1) \setminus S(x,n)$ are divisible by p_n but not by any of $p_1,\ldots,p_{n-1}$. Hence they are of the form kp_n ($\le x$), where k is not divisible by $p_1,\ldots,p_{n-1}$. In other words, k belongs to $S(x/p_n, n-1)$. One checks easily that this remains valid for $n = 1$. □

Combining these two lemmas, we obtain:

Proposition F.3 (Meissel's formula) *Let $\pi(x^{1/3}) = M$ and $\pi(x^{1/2}) = N$. Then*

$$\pi(x) = \phi(x,M) - \sum_{n=M+1}^{N} \pi\left(\frac{x}{p_n}\right) + \tfrac{1}{2}(N-M+1)(M+N-2).$$

Proof By Lemma F.2, we have

$$\phi(x, M) - \phi(x, N) = \sum_{n=M+1}^{N} \phi\left(\frac{x}{p_n}, n-1\right).$$

For n as stated, we have $x^{1/3} < p_n \leq x^{1/2}$, hence $x^{1/2} \leq x/p_n < x^{2/3}$. We can apply Lemma F.1, since $p_n > (x/p_n)^{1/2}$ and $p_{n-1} < x/p_n$. We obtain

$$\phi\left(\frac{x}{p_n}, n-1\right) = \pi\left(\frac{x}{p_n}\right) - n + 2.$$

Also, $\phi(x, N) = \pi(x) - N + 1$. Now

$$\sum_{n=M+1}^{N} (n-2) + (N-1) = \tfrac{1}{2}(N-M+1)(M+N-2).$$

The stated identity follows. □

To use this expression to calculate $\pi(x)$, we need to evaluate a number of terms $\pi(y)$ with $y \leq x^{2/3}$, together with $\phi(x, M)$. So the problem is now to calculate $\phi(x, M)$. As we have seen, this is very easy if $M \leq 3$, but otherwise it is a sieving calculation of length x, which is exactly what we are trying to avoid! But by lemma F.2 again, we can split it as follows:

$$\phi(x, M) = \phi(x, M-1) - \phi(x/p_M, M-1).$$

Each of these terms can be split again, and so on. The terms obtained will always be of the form $\phi(x/n, m)$, representing a sieving calculation of length x/n (which may simply be a case of lemma F.1). A rule is needed to determine whether to perform this calculation as it stands or to split the term again. Different rules have been tried. Lehmer's choice was to continue splitting until $m = 5$ or $x/n < p_m$ (so that $\phi(x/n, m) = 1$ and sieving is completely avoided). However, this strategy can lead to a very large number of terms. Instead, one can treat m and n independently, and agree to stop splitting if either $m = 3$ or $n > n_0$, where n_0 is a chosen value. The first condition could be replaced by $m = 0$. The second condition means that sieving calculations of length up to x/n_0 are allowed. The article [LMO] considers various choices of n_0, giving an analysis of time and storage demands as well as discussing computational procedure. For most cases, it is shown that $x^{1/3}$ is a satisfactory choice for n_0. Later articles by these and other authors have refined the method further.

It is a quite manageable exercise to calculate $\pi(10,000)$ by hand, using no more than a pocket calculator for divisions and the table in appendix G to read off values of $\pi(n)$ for $n < 450$.

Appendix G

Table of primes in residue classes mod 30

Apart from 2, 3 and 5, all prime numbers belong to one of the residue classes (mod 30) represented by 1, 7, 11, 13, 17, 19, 23, 29. The table allots a space to each such number up to 2520, successive blocks of length 30 occupying vertical columns. Numbers are only printed if they are prime; a blank space means that the number belonging there is not prime. This illustrates the distribution of prime numbers between the residue classes, as well as showing up concentrations of primes and gaps.

Table of primes in residue classes mod 30

(1)		31	61			151	181	211	241	271		331
(7)	7	37	67	97	127	157				277	307	337
(11)	11	41	71	101	131		191		251	281	311	
(13)	13	43	73	103		163	193	223		283	313	
(17)	17	47		107	137	167	197	227	257		317	347
(19)	19		79	109	139		199	229				349
(23)	23	53	83	113		173		233	263	293		353
(29)	29	59	89		149	179		239	269			359

(1)			421				541	571	601	631	661	691
(7)	367	397		457	487		547	577	607			
(11)		401	431	461	491	521				641		701
(13)	373		433	463		523			613	643	673	
(17)				467			557	587	617	647	677	
(19)	379	409	439		499				619			709
(23)	383		443		503		563	593		653	683	
(29)	389	419	449	479	509		569	599		659		719

(1)		751		811						991	1021	1051
(7)	727	757	787			877	907	937	967	997		
(11)		761		821		881	911	941	971		1031	1061
(13)	733			823	853	883					1033	1063
(17)			797	827	857	887		947	977			
(19)	739	769		829	859		919			1009	1039	1069
(23)	743	773			863			953	983	1013		
(29)			809	839			929			1019	1049	

(1)				1171	1201	1231		1291	1321		1381	
(7)	1087	1117				1237		1297	1327			
(11)	1091		1151	1181				1301		1361		
(13)	1093	1123	1153		1213			1303				1423
(17)	1097			1187	1217		1277	1307		1367		1427
(19)		1129				1249	1279				1399	1429
(23)	1103		1163	1193	1223		1283			1373		1433
(29)	1109				1229	1259	1289	1319			1409	1439

(1)		1471		1531			1621				1741	
(7)	1447				1567	1597	1627	1657			1747	1777
(11)	1451	1481	1511		1571	1601				1721		
(13)	1453	1483		1543				1663	1693	1723	1753	1783
(17)		1487				1607	1637	1667	1697			1787
(19)	1459	1489		1549	1579	1609		1669	1699		1759	1789
(23)		1493	1523	1553	1583	1613				1733		
(29)		1499		1559		1619			1709			

(1)	1801	1831	1861			1951		2011				2131
(7)			1867				1987	2017				2137
(11)	1811		1871	1901	1931					2081	2111	2141
(13)			1873		1933		1993		2053	2083	2113	2143
(17)		1847	1877	1907			1997	2027		2087		
(19)			1879				1999	2029		2089		
(23)	1823			1913		1973	2003		2063			2153
(29)			1889		1949	1979		2039	2069	2099	2129	

(1)	2161		2221	2251	2281	2311	2341	2371				
(7)					2287		2347	2377		2437	2467	
(11)							2351	2381	2411	2441		
(13)		2203			2293			2383			2473	2503
(17)		2207	2237	2267	2297		2357		2417	2447	2477	
(19)	2179		2239	2269				2389				
(23)		2213	2243	2273		2333		2393	2423			
(29)					2309	2339		2399		2459		

Appendix H
Biographical notes

This appendix contains brief biographical notes on some of the mathematicians who made significant contributions to the material covered in this book. One fact that emerges clearly is the truly international nature of mathematical endeavour! Acknowledgement is due to the historical website of St. Andrews University:

`http://www-history.mcs.st-andrews.ac.uk/history/Mathematicians`

Pafnuty Lvovich Chebyshev, 1821–1894.

Chebyshev (sometimes written Tchebycheff, following the French system of romanization) was foremost among Russian mathematicians of the nineteenth century. He was a student at Moscow University, but spent most of his working life at St. Petersburg. He is regarded as the founder of the St. Petersburg school of mathematics.

He was a highly versatile mathematician who made significant contributions in a number of widely different areas of the subject. His pioneering contribution to the prime number theorem is described in chapters 1 and 2 of this book. Another contribution to number theory was his influential book *Teoria sravneniy* (*The Theory of Congruences*), which appeared in 1849. In probability theory, he gave the world "Chebyshev's inequality" and, more substantially, the "law of large numbers". In the theory of interpolation and approximation, 'Chebyshev sets" have become a definition, and the "Chebyshev polynomials" play a fundamental role. Chebyshev worked in integration theory, for example, generalizing the beta function. He also did original work in mechanics, notably on the problems involved in converting rotary motion to linear motion.

He had a great influence on the development of mathematics in Russia, his pupils including Markov and Lyapunov, among many others.

Peter Gustav Lejeune Dirichlet, 1805–1859

Dirichlet grew up in the German Rhineland, in a family of mixed French and German origins. While he was studying in Cologne, his interests turned to mathematics, against his father's wish for him to become a merchant. He continued his studies at Paris, then pre-eminent in the world of mathematics under the leadership of Laplace and Legendre. He managed to acquire one of the few available copies of Gauss's classic *Disquisitiones Arithmeticae*, which contained an enormous amount of new mathematics, but was presented in a style that other mathematicians found hard to penetrate. Dirichlet treasured his copy and made it his mission to understand Gauss's work and to pass on his understanding to others.

He established his reputation by giving a proof of the case $n = 5$ of Fermat's so-called "last theorem". This led to appointments as professor in Berlin, at the early age of 26, and eventually as the successor of Gauss at Göttingen in 1855.

Dirichlet's most famous contribution to number theory is the theorem described in chapter 4, published in 1837. Even in this short book we have seen several other examples of his work and ideas, such as the notion of "Dirichlet series", and he went on to produce pioneering work on quadratic forms and number fields. Both his lectures and his books were regarded as a model of clear presentation. His work on number theory was eventually summarized in a highly influential book, *Vorlesungen über die Zahlentheorie*, which was only published after his death.

Dirichlet also made important contributions in the area of partial differential equations and potential theory, where his name is attached to the "Dirichlet problem".

Paul Erdös, 1913–1996

Paul Erdös was born into a Hungarian Jewish family in Budapest. Both parents were teachers of mathematics, and it soon became clear that Paul had inherited their interest. He entered university in 1930, and within four years had completed his doctoral thesis on primes in intervals. He was given a position at Manchester, and then at Princeton. When this expired, it was the beginning of a lifetime of constant travel and visiting. He never held a conventional university teaching job. He lived out of suitcases and called himelf a citizen of the world. He had a language all his own: for example, children were "epsilons", men were "slaves" and women were "bosses"!

He was completely devoted to mathematics and surpassed even Euler's output by writing no fewer than 1500 papers in his lifetime. Many of them were written in collaboration, which has led to the game of "Erdös numbers" for mathematicians. This number is the length of the shortest chain of joint authorships leading to Erdös, so that those who have written a joint paper with Erdös have number 1, those who have a joint paper with a 1 have number 2, and so on (on the strength of one paper in a very different field, the author is a rather lucky 2).

Erdös's work ranged through most areas of mathematics, but his real specialities were number theory, graph theory and combinatorics. His most famous achievement in number theory was the "elementary" proof of the prime number theorem, published simultaneously (but not jointly) with Selberg in 1949.

Leonhard Euler, 1707–1783

Euler grew up near Basel, Switzerland. His father, a Calvinist pastor, wanted him to study theology, and Leonhard entered Basel University at the age of 13. His interests moved increasingly to mathematics. His father could partly blame himself, having introduced his son to the subject. Although Leonhard did not go into the ministry, he retained his father's strong religious faith throughout his life. At Basel, he benefited from private lessons from Jean Bernoulli, a leading mathematician of the day.

The Imperial Academy of St. Petersburg had been founded in 1725 by Peter the Great, with the intention of attracting leading scholars to Russia. Nicolaus and Daniel Bernoulli, sons of Jean, were among those appointed, and at their recommendation Euler was invited to the academy in 1727. The rest of his life was divided between St. Petersburg (1727–41 and again after 1766) and the Royal Academy in Berlin (1741–66), to which he was invited by Frederick the Great of Prussia. At that time, these academies offered more favourable conditions for research than universities.

Euler devoted himself incessantly to research throughout his life. In quantity, his output has few rivals in the entire history of mathematics. More than 500 of his papers and books appeared during his lifetime, and a further 400 were published after his death. The task of historians has been made more difficult by his habit of allowing his papers to accumulate in a pile, with the result that they were often published in the wrong order. He became blind in 1769, but he had an extraordinary memory and continued the work unabated, dictating his papers to his sons.

Euler made important contributions to all the areas of mathematics then known. Several of them appear in this book. Other achievements in number theory included giving proofs of several of Fermat's famous unproved assertions. Much of the notation we now take for granted was established by Euler, such as π, i, e^x, $\log x$. His work on infinite series and products led to the relation $e^{ix} = \cos x + i \sin x$ and the product expression for $\sin x$ as well as the "Euler product". His work on the vibration of strings opened up the study of partial differential equations. He made great advances in the dynamics of rigid bodies, introducing the concept of moment of inertia, and laid the foundations of fluid mechanics and calculus of variations. In a completely different field, he paved the way to modern graph theory with his demonstration of the impossibility of walking a circuit of the bridges of Königsberg without duplication.

Carl Friedrich Gauss, 1777–1855

Without any question, Gauss was one of the greatest mathematicians of all time – some would say the greatest. He was born into a German family of limited education, but his exceptional gifts soon became apparent in his childhood. It is reported that when he had only just started going to school, the teacher set the class the exercise of adding the numbers from 1 to 100. The child Gauss worked out for himself the principle of an arithmetic series and amazed the teacher by giving an instant correct solution.

By a fortunate chain of circumstances, Gauss's talent became known, and his royal ruler, the Duke of Brunswick, granted finance for university studies, in the face of opposition from his father. Gauss's doctoral thesis, finished in 1799, was already a landmark in mathematics. It was the first correct proof of the "fundamental theorem of algebra", the statement that every complex polynomial has (complex) zeros. The leading mathematicians of the time had been trying to do this for more than 50 years, but had not been able to produce a satisfactory proof.

At the age of 24, Gauss produced his *Disquisitiones Arithmeticae*, a masterpiece that laid the foundations for the further development of number theory. Here, for example, the idea of congruence was introduced into the subject. Gauss summarized known results and added many more of his own, such as the famous quadratic reciprocity theorem. However, his style was concise, even cryptic, and made no concessions to motivation. It was left to Dirichlet (q.v.) to make the *Disquisitiones* more generally understood.

In these early days, Gauss kept a diary in which he noted his discoveries for his own satisfaction. Some of these discoveries were never published and were eventally rediscovered by other mathematicians. The style of these entries is exemplified by:

$$\text{num} = \Delta + \Delta + \Delta,$$

by which Gauss meant that he had shown that every integer can be expressed as the sum of three triangular numbers.

After the minor planet Ceres had been observed for a short period in 1801, Gauss performed elaborate calculations predicting when and where it would become visible again. His predictions proved to be highly accurate a year later. This success brought him instant fame and a position as director of the Göttingen observatory, which he retained for the rest of his life. This study led to Gauss's next masterpiece, *Theoria motus corporum coelestium*, published in 1809.

There is hardly an area of mathematics – or physics – where one can escape the name of Gauss! He founded the study of electromagnetism, in which the "gauss" is a unit. This study depended on vector calculus, in which Gauss's "divergence theorem" is a fundamental result. "Gaussian elimination" is a basic process in linear algebra. His work involved him in practical observation, both in astronomy and in the earth's magnetism, and this led to an interest in the error of observations, and thereby to probability theory and the notion of the "Gaussian" (alias "normal") distribution.

Jacques Hadamard, 1865–1963

Hadamard was born at Versailles. Both his parents were teachers, and they encouraged him in his mathematical studies at Paris. His doctoral thesis, finished in 1892, was on complex function theory, after which he took up a teaching appointment in Bordeaux. Only four years later, Hadamard achieved his spectacular success with the prime number theorem.

His working life continued for many years after this – and his natural life for many more. In total he published more than 300 papers, on a wide variety of topics. He made important contributions to complex analysis, where the "three circles" theorem bears his name. Much of his work was concerned with analytic continuation and singularities of complex functions. He was a pioneer in considering functions of "variables" that are themselves functions, opening up the whole subject area now known as functional analysis.

Hadamard held a succession of Chairs in Paris from 1909 until retiring in 1937. During World War II he went first to the United States and then to Britain, where (now in his late seventies) he took part in operational research for the Royal Air Force. He returned to France in 1945. After retiring for the second time, he lived peacefully to the age of 97.

Franz Mertens, 1840–1927

Franz Mertens was born in an area of Prussia that became part of Poland in 1919. He studied in Berlin, then a leading centre of mathematical activity under Weierstrass, Kronecker and Kummer. Mertens wrote a doctoral dissertation on potential theory, but his long and productive mathematical life was devoted mainly to analytic number theory and algebra. He moved on to appointments at Kraków (1865), Graz (1884) and Vienna (1894). Several of his contributions appear in this book, notably his estimation of $\sum_{p\in P[x]}(1/p)$, his theorem on the multiplication of series and his proof that $\zeta(1+it)\neq 0$ (which has been used for this and related results ever since). Also, he greatly simplified Dirichlet's original proof of the theorem on primes in residue classes. The "Mertens conjecture", stating that $|M(x)|\leq x^{1/2}$, was formulated in 1897, but the Dutch mathematician T. J. Stieltjes had in fact already claimed to have a proof of it (with an intervening constant) in 1885; needless to say, Stieltjes' claim was not substantiated.

August Ferdinand Möbius, 1790–1868

Möbius was born, and lived for most of his life, in the Kingdom of Saxony, one of the largest of the independent states that later joined to form united Germany. He was educated at home until he was 13. He entered the University of Leipzig to study law, but soon switched to mathematics, astronomy and physics. The astronomer Mollweide was a strong influence during this period. Möbius continued his studies at Göttingen, under Gauss, and at Halle, under Pfaff. His doctoral thesis, completed in 1815, was on the occultation of fixed stars. Soon after this, at the early age of 26, he was given a position as "extraordinary professor" in astronomy and higher mechanics at his original university, Leipzig. He stayed there for the rest of his working life, becoming a full professor in 1844 and (like Gauss) director of the observatory in 1848.

Möbius continued to work productively in both astronomy and mathematics throughout his career. Even within mathematics, he made contributions in widely different areas. In this book, we have met the Möbius function and inversion formula. He published influential books on analytical geometry and statics. "Möbius transformations" have become a standard notion in complex analysis, and the "Möbius strip" is a well-known idea in topology.

Georg Friedrich Bernhard Riemann, 1826–1866

Riemann was one of six children of a Lutheran pastor in the Kingdom of Hanover, another of the states that subsequently united to form Germany. Despite early promise in mathematics, he entered Göttingen University in 1846 to study theology. With his father's consent, he soon switched to mathematics. Among his teachers were Gauss (at Göttingen) and Dirichlet (at Berlin), both of whom were very impressed. His achievements were such that by 1859 he was appointed professor at Göttingen to succeed Dirichlet. Three years later his health broke down, and he died in 1866 at the early age of 39.

In the course his short working life, he made fundamental contributions to several areas of mathematics, all distinguished by an exceptional streak of originality. His doctoral thesis (1851) introduced the notion of "Riemann surface" into complex analysis, initiating the study of curves and surfaces that was later formalized in topology. In 1854, he had to give an inaugural lecture, required even for the junior position of *Privatdozent*. He offered three possible topics, from which Gauss chose the foundations of geometry, a notoriously difficult subject. Riemann's lecture is regarded as a classic milestone in the history of mathematics. It opened up the whole subject of non-euclidean geometry, generalizing from the idea of geometry on the surface of a sphere. These ideas were used in Einstein's theory of relativity. His famous paper of 1859, setting out an agenda for the prime number theorem based on what we now call the "Riemann zeta function", was his only publication in number theory. As we have seen, it formulated the "Riemann hypothesis", which is recognized to this day as one of the most important unsolved problems in mathematics.

Other theorems and concepts bearing his name are to be found in a number of different areas. The reader will have met the Riemann integral, which (while less revolutionary than the concepts mentioned above) remains a core notion of mathematical analysis. We have met the Riemann-Lebesgue lemma in this book. The Riemann mapping theorem is a superb result in complex analysis, and his name appears again in the theory of partial differential equations.

Atle Selberg, 1917–

Selberg was born in Langesund, Norway. His special interest in mathematics developed while he was still at school. He was particularly intrigued by the life and works of the Indian mathematician Ramanujan, who made striking contributions to number theory during a few short years at Cambridge before a tragic early death in 1920. Selberg was a student at Oslo and stayed there as a research fellow until 1947. He was then invited to the Institute of Advanced Study at Princeton in the United States, where he has remained, apart from short breaks, to this day.

His most widely known achievement is probably the "elementary" proof of the prime number theorem, published simultaneously (but not jointly) with Erdös in 1949. However, Selberg has made other highly influential contributions to number theory throughout a long working life. For example, he proved that at least a positive proportion (in a certain sense) of the zeros of the zeta function lie on the line Re $s = \frac{1}{2}$, and his profound development of sieve methods has found many uses.

Charles-Jean de la Vallée Poussin, 1867–1962

Charles de la Vallée Poussin lived in the university town of Louvain, Belgium, throughout his long life. His father was professor of geology there. Charles started the training course for Jesuits but withdrew from it to study engineering. After completing this course he became increasingly interested in mathematics. He became an assistant at the university in 1891 and only two years later, at the age of 26, was appointed to the chair. He held this position for more than 50 years.

His best known contribution is undoubtedly his proof of the prime number theorem, which appeared in 1896, quite independently of Hadamard's proof. He continued his studies of the zeta function, but also did important work on polynomial approximations, complex analysis and potential theory. His *Cours d'analyse* was highly influential; it was revised and updated several times during his lifetime. A second book on potential theory was held up by the 1939–45 war and only appeared in 1949.

Bibliography

[Ap] T. M. Apostol, *Introduction to Analytic Number Theory*, Springer (1976).

[BD] P. T. Bateman and H. G. Diamond, A hundred years of prime numbers, *Amer. Math. Monthly* 103 (1996), 729–741.

[Dav] H. Davenport, *Multiplicative Number Theory*, 2nd ed., Springer (1980).

[Dia] H. G. Diamond, Elementary methods in the study of the distribution of prime numbers, *Bull. Amer. Math. Soc.* **7** (1982), 553–589.

[Edw] H. M. Edwards, *Riemann's Zeta Function*, Academic Press (1974).

[Ell] W. Ellison and F. Ellison, *Prime Numbers*, Wiley (1985).

[Est] T. Estermann, *Introduction to Modern Prime Number Theory*, Cambridge Univ. Press (1961).

[GK] S. W. Graham and G. Kolesnik, *Van der Corput's Method of Exponential Sums*, London Math. Soc. Lecture Notes 126, Cambridge Univ. Press (1991)

[H&Wr] G. H. Hardy and E. M. Wright, *An Introduction to the Theory of Numbers*, 5th ed., Oxford Univ. Press (1979).

[HST] E. Hlawka, J. Schoissengeier and R. Taschner, *Geometric and Analytic Number Theory*, Springer (1991).

[Ing] A. E. Ingham, *The Distribution of Prime Numbers*, Cambridge Univ. Press (1932).

[Ivić] A. Ivić, *The Riemann Zeta Function*, Wiley (1985).

[LMO] J. C. Lagarias, V. S. Miller and A. M. Odlyzko, Computing $\pi(x)$: the Meissel-Lehmer method, *Math. Comp.* **44** (1985), 537–560.

[Lev] N. Levinson, A motivated account of an elementary proof of the prime number theorem, *American Math. Monthly* **76** (1969), 225–245.

[Nar] W. Narkiewicz, *The Development of Prime Number Theory*, Springer (2000).

[Nath] M. B. Nathanson, *Elementary Methods in Number Theory*, Springer (2000).

[New] D. J. Newman, *Analytic Number Theory*, Springer (1998).

[Patt] S. J. Patterson, *An Introduction to the Theory of the Riemann Zeta-Function*, Cambridge Univ. Press (1988).

[Ten] G. Tenenbaum, *Introduction to Analytic and Probabilistic Number Theory*, Cambridge Univ. Press (1995).

[T&MF] G. Tenenbaum and M. Mendès France, *The Prime Numbers and Their Distribution*, American Math. Soc. (2000).

[Titch] E. C. Titchmarsh, *The Theory of the Riemann Zeta-Function*, 2nd ed., Oxford Univ. Press (1986).

[Widd] D. V. Widder, *An Introduction to Transform Theory*, Academic Press (1971).

Index